Western Oregon Wildfires:
An Anthology

REMEMBER ARNOLD VON WINKELREID.

WESTERN OREGON WILDFIRES: AN ANTHOLOGY

By Dr. Bob Zybach

PUBLISHED BY NW MAPS CO. COTTAGE GROVE, OREGON USA

Cover and title page photo of author by Nana Lapham, NW Maps Co., Abbot Butte, B&B Complex Fire, May 16, 2004.

CITE: Zybach, Bob 2024. *Western Oregon Wildfires: An Anthology*. NW Maps Co., Cottage Grove, Oregon USA: 132 pages. ISBN 978-1-7321276-8-5

ACKNOWLEDGEMENTS

This anthology is a collection of articles, editorials, reports, and interviews that I have written, given, or co-authored regarding western Oregon wildfire history and mitigation during the past 20 years, from 1994 through 2023. This is a topic I have researched and worked with my entire adult life, and really began with deer hunting and tree planting in the Tillamook Burn during my high school years in the early 1960s.

As with my earlier anthology in this series, *The Elliott*, special thanks are due Cristy Rein for allowing me to retain the original content and formatting for these articles. April 2024 will mark the 45th year that she has successfully published and edited *Oregon Fish & Wildlife Journal,* which has a circulation of 10,000 mostly rural Oregon residents and businesses -- and also including all elected officials in Oregon and eight other western states, and members of the Executive Cabinet and Congress in Washington, DC.

Thanks are also due Jim Petersen and *Evergreen Magazine* for permission to reprint his 1994 interview with me. Jim also took the photo of me for the cover of that issue (above) -- hard to believe that was 30 years ago!

TO THE READER:

This is the second in a projected series of four anthologies of my published articles, reports, interviews, and editorials from the past 40+ years. The previous collection focused on Elliott State Forest history, management, wildlife, and politics; this book focuses on wildfire history and mitigation of western Oregon forests. Subsequent anthologies will focus on Pacific Northwest forest reforestation and restoration; and on spotted owls, coho, their designated critical habitats, and the Endangered Species Act. These works are written from the perspective of a forest scientist and historian with strong opinions, based on extensive personal research and experience.

The collected articles contain significant scientific data in the forms of maps, tables, historical and documentary photographs, and the writings and spoken words of recognized experts -- but they have not been "peer reviewed" in the academic sense as that term is currently used; rather, the intended audience for these writings are my actual peers and reviewers: loggers, foresters, tree planters, roadbuilders, hunters, hikers, and fishermen. Rural businesses, residents, schoolkids, and grandparents. The people who have always lived, and worked, and played in the Elliott and in other western forests. True indigenous knowledge.

This intention of this series is to present a selection of works on a given topic in the format in which they were first published, including typos, ads, and even outdated or possibly incorrect information. The intended result is to retain original context and ambiance so much as possible.

Selections are mostly arranged chronologically, with the 1994 to 2010 timeframe including a commissioned report, two articles, an editorial, and an interview with Jim Petersen of *Evergreen Magazine.* The post-2010 timeframe is comprised of more than a dozen article/editorials written for *Oregon Fish & Wildlife Journal* from 2012 to 2023, as formatted by editor Cristy Rein.

Other collaborations and publishers:

-- The report on Elliott fire history (p.2) was commissioned by Oregon Forest Industries Council, Associated Oregon Loggers, Northwest Forestry Association, and Douglas Timber Operators, and written under the guidance of Greg Miller.

-- The article on Northwest forest fire history (p. 32) was coauthored with Jennifer Solomon, editor for *Forests Today and Forever,* a magazine published by the Oregon Small Woodlands Association.

-- The wildfire economics article (p. 37) was coauthored by John Marker, Mike Dubrasich, and Greg Brenner for the US Department of Agriculture *Wildland Fire Lessons Learned Center* website; this is the condensed online version of the original 22-page HTML report. John also coauthored a print article from this material (p. 60) that was published four years later.

Bob Zybach

Creswell, Oregon

January 22, 2024

WESTERN OREGON WILDFIRES: AN ANTHOLOGY

TABLE OF CONTENTS

Michael Cheetham
November 19, 2018

FILE UNDER: "A WARNING THAT WENT UNHEEDED" . . . OR . . . "PREVENTABLE INFERNOS"

IN CASE YOU MISSED IT -- on August 8th of this year, the following article appeared in "The Daily Caller." It contained a well-informed and sadly prescient warning from forestry expert, Dr. Bob Zybach.

READ -- and be righteously INFURIATED.

"DECADES OF MISMANAGEMENT TURNED U.S. FORESTS INTO 'SLOW-MOTION TIME BOMBS' "

by Michael Bastasch | Energy Editor
"The Daily Caller"
August 8, 2018

- Wildfire experts say poor management, not global warming, is the major reason behind worsening wildfires.
- Forester Bob Zybach warned decades ago that environmental regulations and less logging would make fires worse.
- Trump administration is doing more active management of lands, but is it enough?

Bob Zybach feels like a broken record. Decades ago he warned government officials allowing Oregon's forests to grow unchecked by proper management would result in catastrophic wildfires.

While some want to blame global warming for the uptick in catastrophic wildfires, Zybach said a change in forest management policies is the main reason Americans are seeing a return to more intense fires, particularly in the Pacific Northwest and California where millions of acres of protected forests stand.

"We knew exactly what would happen if we just walked away," Zybach, an experienced forester with a PhD in environmental science, told The Daily Caller News Foundation.

Zybach spent two decades as a reforestation contractor before heading to graduate school in the 1990s. Then the Clinton administration in 1994 introduced its plan to protect old growth trees and spotted owls by strictly limiting logging.

Less logging also meant government foresters weren't doing as much active management of forests — thinnings, prescribed burns and

Guest Editorial: The Coming Firestorms

By Dr. Bob Zybach

The most deadly, destructive, and widespread catastrophic-scale forest fires in Oregon's history erupted on Labor Day 2020, driven by strong east winds.

Unless we change how our national and state forests are managed, these events will be just one more chapter in this age of predictable, increasing, and ever-greater firestorms.

I've spent my career studying forest fires and forest health. In a 2018 Daily Caller interview, a few weeks before the California Camp Fire destroyed the town of Paradise, I said: "You take away logging, grazing and maintenance, and you get firebombs." Then someone took my quote, put it on a forest fire photo, and posted it from the ruins of Paradise. The resulting meme quickly went viral on Facebook.

This September, Facebook began flagging this post as "partly false" because my quote, and related interview, didn't mention climate change. That is because my documented predictions, based on significant research and personal experience, do not consider changing climate, in order to be accurate.

The broad arc of Oregon's fire history explains why this year's catastrophic wildfires have converted our public forests into unprecedented firebombs. What were once green trees filled with water, have now become massive stands of pitchy, air-dried firewood.

For thousands of years ancestral Indian families kept ridgeline and riparian areas open for travel, hunting, fishing, and harvesting purposes. They cleared ground fuels by constant firewood gathering, root harvesting, and seasonal fires.

These actions created widespread systematic firebreaks in a beautiful landscape characterized by foot trails, grass prairies, southern balds, huckleberry fields, camas meadows, oak savannah, and islands of mostly even-aged conifers.

Following the historic 1910 firestorms, the US Forest Service established a nationwide network of fire lookouts and pack trails backed up by rapid response fire suppression. This system became remarkably effective over time.

From 1952 until 1987, for 35 years, only one forest fire in all of western Oregon was greater than 10,000 acres: the 1966 43,000-acre Oxbow Fire in Lane County.

But since 1987, the past 34 years, Oregon has had more than 30 such fires, with several larger than 100,000 acres.

The 2020 Labor Day Fires alone covered more than one million acres, destroyed over 4,000 homes, caused 40,000 emergency evacuations, killed millions of wild animals, and thickly blanketed the state with an acrid, unsightly and unhealthy smoke for nearly two weeks.

What changed to cause this dramatic increase in catastrophic wildfire frequency and severity?

The problems began in the 1960s, with apparently well-intentioned national efforts to create large untouchable wilderness areas and cleaner air and water on our public lands.

The single biggest turning point in how public forests are managed happened on December 22, 1969: about 50 lawyers in Washington, DC created the Environmental Law Institute, and a short distance away Congress simultaneously passed the National Environmental Protection Act (NEPA).

Next, the 1973 Endangered Species Act (ESA) and the 1980 Equal Access to Justice Act (EAJA) provided the growing environmental law industry with a way to be paid by the government for challenging nearly every attempt to log or actively manage public forests.

By the 1980s, the artificial creation of Habitat Conservation Plans ("HCPs") and the listing of spotted owls as an Endangered Species laid the groundwork for today's fires.

The 1994 Clinton Plan for Northwest Forests might have been the final nail in the coffin. The subsequent never-ending environmental lawsuits, new Wilderness areas and HCP creations, access road decommissionings, and fruitless public planning exercises have created tens of millions of acres of massive fuel build-ups and "let it burn" policies that have decimated our forests and wildlife.

A predicted result has been ever larger western Oregon forest fires. More than 90% of these large and catastrophic scale fires have taken place in federal forestlands, which represent almost 60% of all Oregon's forested areas.

Even if, like Facebook executives, you believe these fires were somehow sparked by climate change, you should be very concerned with what will happen next.

Lessons from the 1902-1929 Yacolt Fires, 1933-1951 "Six-Year Jinx" Tillamook Fires, and the 1987-2018 Kalmiopsis Wilderness Fires are clear: unless removed, the dead trees resulting from these fires will fuel even greater and more severe future fires.

The 2020 fire-killed trees should be strategically mapped, sold, and harvested ASAP, before they further deteriorate in value and increase in risk. Prices for Douglas fir logs are at record highs, and there is great current need for good-paying rural jobs and local building materials.

It will be interesting to see if we can learn from Oregon's fire history and take the prompt, decisive actions needed to avoid the clearly predictable coming firestorms.

Bob Zybach has been Program Manager for educational nonprofit www.ORWW.org since 1996. He is author of The Great Fires: Indian Burning and Catastrophic Forest Fire Patterns of the Oregon Coast Range, 1491-1951.

Use of Fire History in the Development of the 1993 Elliott State Forest Management Plan Draft

By
Bob Zybach
Corvallis, Oregon

February 14, 1994

Prepared For
Greg Miller, Director of State Land Purchasers Division
Oregon Forest Industries Council
Associated Oregon Industries
Salem, Oregon

Introduction

This review of the Elliott State Forest Draft Plan ("the plan") has been completed under written contract for the State Timber Purchasers Division of the Oregon Forest Industries Council (OFIC) at the request of its director, Greg Miller.

In particular, OFIC requested:

1. An analysis of Oregon Department of Forestry's (ODF) "use of *fire history* for developing their recommended alternative," Strategy 6;
2. A description of "how the *natural fire cycle* has shaped" the Elliott State Forest;
3. An assessment, based upon consideration of the plan's given fire cycles and fire history, of any *"potential for shorter rotations* to obtain the *same wildlife objectives* (owls and murrelets)" given in Strategy 6.

The following discussion outline was used to address the three OFIC requests:

A. Current forest structure and 300-year fire history of the Elliott State Forest.
B. Definition of "natural fire cycle."
C. Forest fires, human population rates, and wildlife habitat dynamics.
D. Historical environments, ESA strategies, and coastal bird populations.
E. Catastrophic fire history, forest productivity, and log income potentials.
F. Conclusions.

The text of the discussion outline is contained in the following pages. Based upon an interpretation of this material, a brief conclusion to each of the OFIC questions can be summarized:

———ODF apparently did not use the 1770 to 1951 fire history described by Phillips in its development of plan alternatives, including the preferred "Strategy 6." Rather, a 150-year fire cycle model— generally ascribed to Agee—seems to have been used, in part to help justify a recent change in management focus from timber production to maintenance and creation of older forest types and conditions. *A fire history perspective would have allowed for far greater latitude in the spatial and temporal designing of logging plans, wildlife habitat creation and maintenance strategies, fire control options, and net income production.*

———The "natural fire cycle" of the Elliott supposes a mathematical predictive model that is biased against human activity (including logging, burning, and road building) and toward "average" decadent stand conditions, numerous older trees, and random lightning strikes. The result of using fire cycle models for planning purposes remains controversial. The "shaping" of Elliott State Forest by fire can be better characterized *historically*, as *the result of a long-term series of botanical responses to constant human disturbances caused by daily and seasonal fires of varying degree and intensity.* This has resulted in an extensive stand of young, even-aged Douglas-fir trees that has persisted for over 120 years; it is quite possible similar vegetation patterns have existed for similar lengths of time in this area during the past few thousand years.

———If the "wildlife objectives" in Strategy 6 are interpreted to mean population maintenance or increase over time, *the mobility of owls and murrelets and their proven resiliency to catastrophic fires would indicate that historical rates of clearcutting can probably be maintained indefinitely* (but past fragmentation caused by this practice should probably be mitigated first); If the "wildlife objectives" are to simply create or maintain certain structural stand characteristics ("desired or required" habitat) within the forest, then these human definitions of desired future conditions would include (and require) a Strategy 6-type approach. The difference in these two positions is the difference between mathematical projections based upon perceived "averages" and interpretive projections based upon documented evidence of disturbance and resilient recovery.

A. Current forest structure and 300-year fire history of the Elliott State Forest.

The Elliott State Forest is approximately 93,000 acres in size, of which about 90% burned in the 1868 Millicoma Fire. This fire killed over 300,000 acres of trees in a few weeks, and is considered one of *the "Great Fires" of Oregon Coast Range history—catastrophic fires of great intensity that covered hundreds of thousands of acres of forestland at a time*, often in a matter of only a few hours or days [Zybach, 1988]. By 1900, most of the burn had reforested to stands of 10 to 30-year old Douglas-fir trees. By 1922, many of these stands were "eight to twelve inches in diameter on the stump" [ODF, 1993: I-5]. Today, the Elliott contains about two billion feet of mature timber, the majority of which is contained in the 100-120 year old stands resulting from the 1868 fire. In the past twenty years over one billion feet of timber has been logged from about 36,000 acres of these fire-established trees [ODF, 1993: I-7]; these last numbers represent about 40% of the forest's total area, but less than 33% of the commercial timber volume that has grown since 1900 (or maybe even 1925).

The fire history of the Elliott is well presented in the plan by the acknowledged expert on the topic, Jerry Phillips. *The question is: how that history was used (if at all) in the development of the preferred alternative.* A clear set of assumptions in the plan is that: 1) long-term school income; 2) general forest health; 3) native biodiversity; 4) spotted owl and marbled murrelet populations; and 5) anadromous fish runs; can all be maintained or improved by attempting to mimic *"natural patterns"* [ODF, 1993: III-28]. Because those patterns—as clearly described and mapped by Phillips [ODF, 1993: I-2-5]—*were primarily a function of forest fire history,* we can look there first for the information needed to best identify and describe the processes and populations considered to be "natural" and desirable to the planning area.

1694. 300 years ago the area that is now the Elliott State Forest was growing many of the fuels that were consumed in the 1770, 1840, and 1868 fires mapped by Phillips. Few of these trees were ever logged. This point in time was nearly four or five human generations before the first European influences began affecting local people, plants, and animal populations. *The forests at that time must have most nearly approximated the "naturally functioning ecosystems" envisioned by government wildlife biologists and old-growth ecologists today.*

What information is available to tell us of environmental conditions at that time? Probably the six best (cheap and accessable) sources are the stumps, snags, and trees remaining on the forest from seedlings established before 1694; the General Land Office (GLO) original land survey notes from the mid-1800s; the rings of bored trees and sound stumps and snags; the timber cruises of the State Forester and his staff between 1915 and 1955; living memory (especially focused oral histories); and photographs (both aerial and terrestrial). The diameter, distribution, and species of the forest's oldest known trees and stands (and their documented remains) can then be sytematically mapped using these sources of information. According to the plan [ODF, 1993: I-3], the 300,000 acre fire of 1868 included 10 to 100-year old stands created after the 1840 and 1770 fires, but "some of [the trees in the area of the fire were] estimated to have been about 300 years old." Were all of the snags measured for age created by the 1868 fire, or could they have resulted from the 1840 or 1770 fires? Or from some other event entirely? The variety and extent of those 300 year old trees need to be known before too strong a commitment is made toward managing for their perceived structural characteristics [ODF, 1993: III-31] over too large an area .

It is reasonable to assume that *most (perhaps as much as 90 or 95%) of the Elliott may have been even-aged stands of juvenile and second growth Douglas-fir 125 years of age and younger in 1694.* That would be very similar—with the exception of the "cookie cutter" design of many post-World War II clearcuts—to the structural and biological forest conditions that exist today. Such an assumtption could be based on the following facts:

1) the 1868 fire was able to burn 90% of the current area of the Elliott [ODF, 1993: I-3];

2) the oldest known stands were then apparently about 300 years old (i.e.; created about 1568);

3) the current land base supports 97% of the acreage in juvenile (also, "precommercial" or "reprod"—between one and about 30 years of age) and second growth (younger than 180 to 220 years of age) stands of Douglas-fir; and

4) portions of the 1770 and 1840 fires also occurred within the current planning bounds (it is possible—and even likely—that the primary fuels of the earlier fires also dated to about 1568, or later).

1770. *Phillip's discovery and mapping of the 1770 Coos fire is significant for the evidence it provides of Indian-caused forest fires that precede European and American contact* [Zybach, 1988: Appendix B: 15-21]. The boundaries of this fire are, due to age, only approximate. The fire seems to have been started by the harvesting and maintenance of an Indian prairie complex along the Umpqua and/or Coquille, Coos, and Millicoma Rivers, although Kalapuyan burning in the Umpqua basin seems the likeliest possibility. The significance of this fire is that it could not have been started by white explorers, settlers or loggers; and that it was probably not started "naturally," (by lightning, volcanic eruption, or some other non-human method) either. So far as currently possible, it seems to demonstrate the probability that the aboriginal human populations of the Oregon Coast Range—and particularly the Kalapuyan families that lived along the eastern rivers of the range— were capabable of starting forest fires that occasionally killed hundreds of thousands of acres of trees at a time.

The existence of this fire also indicates that the "natural pattern" of the western slope of the Oregon Coast Range may have been dominated by nearly solid canopies of even-aged stands of mostly Douglas-fir, stretching from the narrow Pacific fog belt of pine, spruce and hemlock on the west, to the vast oak and grass savannahs of the eastern part of the range. This cover would be mottled and bounded with strips and patches of even-aged alder (becoming more common along marshes, seeps, creeks, and landslides to the west), scattered prairie complexes (concentrated along the edges of estuaries, rivers, and large creeks), meadows (flats, ridgelines, and southern slopes), balds (peaks), and berry patches, and by large areas of snags, burned ground, and/or juvenile Douglas-fir trees (often located to the southwest of the prairies and meadows; a likely combination of seasonal fires and seasonal east winds).

1840. In 1840, missionary Gustavus Hines made an eyewitness account of a forest fire taking place to the immediate northeast of the Elliott, near present-day Elkton. His account is remarkable for four reasons: first, it seems to be the first historical record available regarding a forest fire in the Oregon Coast Range; second, it clearly demonstrates the potential for Indian campfires or prairie fires along the eastern boundary of the Coast Range to enter the forests to the west; third, *this event seems to herald the beginning of three decades of catastrophic forest fires in western Oregon, peaking in 1849-50, and again in 1867-68;* and finally, this account seems to confirm Phillip's research, including the general direction of the fire, which seems to have started west of present-day Eugene, and traveled southwesterly into the area of the Elliott. This is also the same basic pattern (northeast to the southwest) that is known for the other Great Fires of the Coast Range [Zybach, 1988: 26-28], precisely as Phillips has mapped the 1770 and 1868 events. Because the 1840 journals of Hines are not well-known today, and because they tie in so well with Phillip's research on the Elliott, I have included a somewhat more detailed account than might otherwise seem necessary:

Hines kept a journal of his entire stay as a Methodist missionary in Oregon, later turning the results into a popular and well-known book. In 1840 he travelled with a fellow missionary, Jason Lee, from their religious community near present-day Salem, to a community of Indian families near present-day Reedsport. They traveled south with pack animals, crossing the Willamette River after several miles and joining the established "California Trail" on a southwesterly course toward present-day Elkton. The California Trail was a pack animal and livestock route between British

Fort Vancouver and the beaver, cattle, and horses of the Mexican Sacramento Valley. It was constructed by the British Hudson's Bay Company (HBC) and the American Smith, Jackson, & Sublette Co. beaver hunting expeditions of 1826, 1828, 1829 and 1830. Portions of the route are still followed today by parts of I-5, Highway 99W, and Territorial Road.

> Friday August 21, 1840 After traveling about twenty miles over a rolling country, presenting almost every variety of scenery, we halted for dinner on a small stream called "Bridge river," [HBC name for Smith River since 1828, at least. The river was possibly the Siuslaw, rather than the Smith] on account of a log bridge having been thrown across it, by some California party. This stream runs in a deep cut, and, but for the bridge, would have been difficult to cross. In the afternoon we passed over the mountain "La Beache," [these are the 1826 Elk Mountains of David Douglas; French Canadian trappers employed by the HBC called elk "la biche," as in Lake LaBish, near present-day Salem. Today we call this mountainous divide between the southern Willamette and northern Umpqua Valleys the Calapooia Mountains] which consists of a vast assemblage of hills thrown together in a wild confusion, and covered with a heavy forest of fir and cedar trees. The latter is the most stately and majestic timber of the kind that I have ever seen. Some of the trees are from ten to fifteen feet in diameter, and towering to an incredible height. On beholding them, one is reminded of the scripture account of the cedars of Lebanon. It required three hours to cross this mountain, and as we descended it to the south, we found the fire making sad havoc with the fine timber with which its sides were adorned. In some places it rages so hard along the trail, that it was quite difficult for us to pass; but, urging our way along, we arrived at sun down at Elk river, and camped on a beautiful plain on its south bank [near present-day Drain, Oregon; according to local resident Gerald Bacon in 1987, this was possibly a grazing area known as "tin pot" to early stock drivers].

After a week of preaching and singing to the Indian families along the lower Umpqua (near present-day Reedsport), Hines and Lee made their return:

> Monday September 1, 1840. . . . In the afternoon, we again passed over the Elk mountain [see "La Beache" mountain note above], and found that the fire was still raging with increasing violence. A vast quantity of the large fir and cedar timber, had been burned down, and in some places the trail was so blockaded with fallen trees, that it was almost impossible to proceed; while now and then we passed a giant cedar, or a mammoth fir, through whose trunk the fire had made a passage, and was still flaming like an oven. Every few moments these majestic spars would come "cracking, crashing, and thundering" to the ground; but while the fire was thus robbing the mountain of its glory, we pushed on over its desolated ridges, and at sun-down arrived on a little prairie at its base, where we made our encampment. Several times during the night we were awakened by the crash of the falling timber, on the mountain, which sometimes produced a noise similar to that of distant thunder.

1868. Phillips has traced the original source of ignition of this event to a settler's clearing fire in the Greenacres area near Scottsburg. His assertion that the fire seems to have mostly occurred in 250 to 300 year old stands of Douglas-fir is based upon his personal identification and measurement of several representative snags and trees within the fire's boundary [personal communication: April 1, 1988]. *Fires were widespread throughout the Douglas-fir Region in 1868,* with an even larger fire complex, the 350,000 to 600,000 acre Yaquina Burn occuring a few miles to the north of the Coos Fire during the same fire season. There were several sources of ignition to the Yaquina Fire, all of them also believed to be human [Zybach, 1988: 100-121].

> . . . records of precipitation at the mouth of the Columbia show 1868 to have had the driest June, July, August, and September in a 58-year period of record for the station. Similarly these months at Vancouver, Washington, were second only to 1866 in having the least rainfall during a 72-year period. [Morris, 1934: 332]

1883. In addition to 1840, another fire year that is poorly documented is 1883. In his first Annual Report as Oregon State Forester, Lynn Cronemiller, describing a portion of the newly-named Elliott State Forest [1930: 20], wrote that: *"The fire of 1868 was undoubtedly extremely severe, for there are very few snags standing in the area*, indicating that practically all the trees were consumed." If it seems unlikely that large, green trees could have entirely vaporized in a single fire, no matter how hot, there is some support for Cronemiller's conjecture. In 1845, an early explorer and journalist named Joel Palmer [1983: 93], for instance, noted that:

> Along the coast from Cape Lookout to the 42d Parallel there is much land that can be cultivated; and even the mountains, when cleared of the heavy bodies of timber with which they are clothed, will be good farming land. There is so much pitch in the timber that it burns very freely; sometimes a green standing tree set on fire will all be consumed; so that it is altogether a mistaken idea that the timber lands of the country can never be cultivated.

At least three other possibilities would seem more likely to explain Cronemiller's observation:

1) that the area had been subjected to repeated cultural fires through time, and therefore did not have a stand of large trees at the time of the 1868 fire (probably the most likely);

2) that the area had burned earlier (possibly during the 1840 fire), and therefore dead, pitchy snags provided a highly—and almost completely—combustable fuel for the 1868 flames;

3) that the area had burned at least one additional time (with an effect similar to the second possibility) between 1868 and Cronemiller's report.

In support of the latter speculation, there are many contemporary references to fires along the Oregon Coast in the 1880's. During the 1883 U.S. Coast and Geodetic Survey that began on August 11 at the mouth of the Umpqua River (about 6 miles west of the Elliott), for instance, it was noted that: *"At the beginning of the season dense smoke from the many forest fires raging along the coast materially impeded the progress of the work . . ."* [1883: 58].

1896. In the 1890s the record of fire in the Elliott becomes much clearer, mostly due to the making and saving of photographs from that time. The photograph on the following page, for instance, shows an 1896 view of the Elkhorn Ranch hunting lodge, an important part of the cultural history of the Elliott. The lodge was owned by the Gould family, who gave this picture to Jerry Phillips sometime before the late 1980s. *The partial results of the 1868 fire are evident in this picture, as is evidence of the presettlelemnt forest and of the current forest.* The Goulds ran sheep, and this may have had the effect of preventing ladder fuels from developing around the base of these snags, and therefore helping to prevent a re-burning of the land, as occured in the 1868 Yaquina Fire, the 1902 Yacolt Fire, and the 1933, 1939, 1945, and 1951 Tillamook Fires.

Another important question: *Was the primary function of these snags in past times as wildlife habitat, or as ephemeral fuels?*

1910. Widespread and deadly catastrophic forest fires in 1902 and 1910 combined with a widespread national fear of an eminent "timber famine" to directly result in a number of local, statewide, and national efforts to "prevent forest fires." This effort initially resulted in the creation of several county fire districts and a focus for the newly-developed state and federal forester's budgets, culminating—at least symbolically—with the creation of Smokey the Bear in the late 1940s. These local and national efforts seem to have been effective (particularly when combined with earlier fire boundaries, clearcuting, broadcast burning, brush control, and road building efforts) in reducing the number and extent of forest fires in the Coast Range. *It could be argued that the advanced age of many of today's trees in the Elliott can trace their existence to grazing practices and fire suppression policies enacted between the late 1800s and the early 1950s.*

1896 view of the Elkhorn Ranch. The partial results of the 1868 fire are evident in this picture, as is evidence of the presettlelemnt forest and of the current forest: *Was the primary function of these snags in past times as wildlife habitat, or as ephemeral fuels?*

B. Definition of "natural fire cycle."

A basic conceptual difference separates two schools of forest history: those who think that only "natural" fires started by nonhuman methods (usually lightning—but also volcanic eruptions, spontaeous combustion, friction, etc.) help define our forests, and those who think that it is "only natural" that people set fires. Throughout the Douglas-fir Region, including Elliott State Forest, historical documentation strongly confirms the latter position and discounts the former. This evidence—in the forms of living memory, tree rings, photographs, maps, etc.— supports the idea that *we live in a forested environment that has been far more affected by human fires over the past several centuries (and probably millennia) than fires from all other causes combined.*

The plan clearly acknowledges that "fire is a basic element that shapes the forest ecosystem." Agee (1991) is quoted as saying that "there is no evidence that [Indians] purposely burned upland forests such as the Elliott," and it is thus concluded that *"wildfires started by lightning* have affected forests in the Elliott area for thousands of years" (ODF, 1993: III-50, emphasis added). Although this conclusion might have some merit, it is unlikely that lightning fires have had anywhere near the influence of human fires, or that "wildfires" even existed at all before European American settlement ("wildfire" in commercial forest areas is a European concept based upon fire suppression practices and, as such, could only exist since the arrival and imposition of European values—less than 200 years ago). Finally, there is no historical record of even a single significant forest fire in the Coast Range that was started by lightning [Zybach, 1988: 31-32]. All of the coast's historical "Great Fires" of the past few centuries have had documented—or highly suspected—human sources of ignition.

It is apparently assumed that because Agee thinks that "no evidence" exists to document local people *purposely burning upland forests*, therefore, lightning did. This reasoning is then extended further to arrive at the cyclic equation that "replacement fires burned through the Elliott area on the average of once every 150 years" (ODF, 1993:III-50; the actual quote says *"fire history studies* suggest that replacement fires burned . . . every 150 years," but Phillip's fire history research details fires through much of the area in 1770, 1840, and 1868—about *once every 50 years* in prehistoric and early historic times). In addition, there are lots of records of purposeful upland forest burning by indigenous people in southwest Oregon [Zybach, 1988: 33 -42]. Besides, whether a timber trespass, fire, or personal injury is "on purpose" or "by accident" is irrelevent—the result is the same and the cause is still "human."

In both prehistoric and historic times, families have existed and moved in systematic fashion across the western Oregon landscape. Thousands of cooking and heating fires have been used daily for thousands of years in the vicinity of Elliott State Forest. On a seasonal basis, people have burned brackenfern, cured tarweed, and cleared forest debris over thousands of acres annually during spring and fall burning seasons. Seasonal east winds, extended periods of drought, a rolling topography, and some of the fastest and largest growing fuels on the planet have combined with these widespread and constant sources of ignition to produce our current forests of fire-dependent and fire-resistant plants. These patterns of plants and cultural fires over time have formed the primary habitat for native wildlife in this region since the first arrival of human hunters and settlers, at least 11,000 years ago.

Because "fire cycle" models are based upon an assumption of random points of ignition and randomly spaced catastrophic events (while people use fire in regular, predictable, and systematic patterns), such models cannot be used to either predict a known past or plan a desired future—at least for most of the forests of the Douglas-fir Region. The historic intervention of people into this area meant that fire was introduced *periodically* (on both daily and seasonal schedules) into the environment for the first time. The difference for local forests was that, rather than evolving through possible *cycles* of predictable conditions (including catastrophic stand replacement fires and various "seral stages"), the tendency has been toward a resilient repopulation of areas periodically burned by fire. As conditions change (a constant series of interrelated processes), the size, shape, appearance, and populations of the forest changes. And, as the forest changes, the timing, size, shape, appearance, and intensity of the fires that periodically enter the forest bounds also change.

A similar, but somewhat different, viewpoint is shared by Pyne:

> The magnitude, character, and organization of these historical changes lead to the concept of a *fuel cyle*. And from this concept comes another, the *fire cycle*. As their names imply, for these concepts historical changes are considered to be regular and at least roughly periodic. In fact, they are neither. In general, only where humans intervene with fire and fuel management practices is there an approximate cycle or an apparent periodicity. . . The fuel cycle and fire cycle concepts are convenient ways to characterize the fact that fuels and fires have histories, but erroneous designations of how those histories evolve and interrelate. (Pyne 1984:103-106)

The significance of the differences between mathematical "fire cycles" and documented "forest histories" to the Elliott plan are critical. If endangered species populations have had to adapt to periodic catastrophic (100,000 acre and larger) forest fires in past centuries, will their long-term survival strategies require some form of return to those past conditions or processes? If so, clearcut logging, slashing, precommercial thinning, and broadcast burning should be considered as important tools in any effort to physically mimic known presettlement forest environments. If not, can "new forestry" practices (that define homogenized "presettlement landscapes" in statistical terms of 50-11-40 equations, snag recruitment strategies, coarse woody debris distribution patterns, and multilayered canopies on an ecosystem-scale basis) successfully mimic past conditions? Would such an effort necessarily result in larger populations of select bird and fish species (including ESA listings, game fish and animals, useful research species, and birds and insects favored for their grace, color, or beauty)? Or would the widespread adoption of these experimental practices create an increasingly volatile and deadly series of fire bombs, as some scientists and analysts suggest (Peterson, 1993: 2)?

The answers to these questions vary widely, depending upon how the historical record is interpreted, or whether statistical value-driven mathematical models are to be trusted. The lesson of history seems to be that we can clearcut, burn, and reforest with near impunity for long periods of time (consider the pasturing and plowing of Greece, Spain, China, or Switzerland). Computer-driven predictive models, on the other hand—based upon assumptions of fire cycles, seral stages, and climax vegetation—seem to say that if we continue clearcutting older trees we will accelerate the extinction process rate (with possible disastrous human consequences) for a number of "late -seral stage" wildlife populations.

C. Forest fires, human population rates, and wildlife habitat dynamics.

The history of catastrophic Oregon Coast Range forest fires is one of incredible, almost over-night changes to vast areas of the physical and biological environment. The Tillamook Fire of 1933, for instance, burned over 220,000 acres of forest in less than one day, sending a 43-mile wide mushroom cloud looming nine miles in the air over the Willamette valley [Zybach, 1982]. As with most other local forest fires of great size [including the 1840, 1868, and 1883 fires), the Tillamook fire occurred during the late summer/early fall east wind "fire season," long after the spring hatch of native birds had learned to fly and fend for themselves. How many birds were trapped in these fires? Or, did birds (and other predators, including humans) flock to the great burns that traced the aftermath of these events, picking off injured and starving rodents, insects, and other exposed prey by the millions? Did the survivors learn to adapt to second growth conditions, as current populations in the Elliott and Tillamook State Forests, and the Siuslaw National Forest,have learned to do? Or did they migrate to other areas and compete for mates and food and shelter with the populations already established in those places? The answers to these questions are very important if we expect to successfully manage for non-game coastal bird species in the years to come.

Today's populations of native coastal birds have all descended from thousands of generations of animals that had to periodically adapt to vastly changed conditions time and time again. Their environment was never a sea of "steady-state" "climax stage", old-growth trees [ODF, 1993: III-31], and never can be. Perhaps it was the process of adapting to periodic fire or wind-caused deforestations over the landscape that helped permit owls and murrelets to survive to the present. Should we then again adopt these processes into the environment? Perhaps even exaggerate their occurrence, in hopes of increasing depleted populations? Or can these effects be simply mimicked, with trees being cut and processed into human products, instead of simply burned and turned into hazardous events and mass air pollution?

It seems to be a matter of the most common sense that *the management of wildlife habitat and native species through time must be based upon established patterns of ecosystem dynamics if the management plan is to be successful.* This problem has been greatly exacerbated over the past several decades due to demands and changes brought about by great increases in local and global human populations. The demands of people upon all of the earth's resources must be seen as the fundamental problem challenging the future of the planet's wildlife populations at this time. The related roles of catastrophic fire, fire suppression activities, and clearcutting must be measured against the expanding *need* for forest products to meet the expanding number of human families being created daily. These human needs (fuel, water, food, shelter, recreation, etc.) are not only derived from forests, but by damming rivers for electricity, planting lawns and other exotic vegetation in our urban areas, growing food crops across great expanses of former prairie lands, and building great freeways of rapidly moving vehicles all across the former migration routes of buffalo, elk, grizzly bears, and antelope. Fire suppression and clearcutting actions have helped to support the increasing human constructions of the past century, but it is debateable how great their impact upon native wildlife habitat has been, compared to other impacts associated with human population growth. These factors need to be especially considered when dealing with flying, swimming, and migrating animals.

D. Historical environments, ESA strategies and coastal bird populations.

A primary concern of logging activities in the Elliott is the harm that might be done to spotted owl and marbled murrelet populations. It is worried that *"habitat removal may cause disturbance extensive enough to disrupt normal bahavior patterns," particularly as those actions might be interpreted through NEPA by the U.S. Fish and Wildlife Service* (USFW)[ODF, 1993: I -22]. The question then, is: What are the normal behavior patterns of these birds, given the Elliott forest's history of young second growth stands of Douglas-fir and over 30,000 acres of clearcutting in the past several decades? A second question is: how does the USFW "interpret" such "actions?" Would the conversion of these stands to an older condition that hasn't occurred for at least 120 years (several dozen bird generations) create an even greater "disruption of normal behavior patterns" than most other management options (including existing management trends)?

The 50-11-40 rule, snag recruitment strategies, and current riparian zone management guidelines all point toward the creation of an idealized, homogenized "old-growth characteristic" condition that has never occurred in the history of the Elliott. Even if the artificial creation of these conditions is physically possible (the 1770 and 1868 fires and the 1962 Columbus Day storm argue heavily against the possibility), *what evidence do we have that this effort will result in greater biodiversity or improve targeted bird populations?* And does any such "evidence" realistically reflect the proven mobility and adaptability of these animals?.

Assuming that the current information regarding the decline in spotted owl and marbled murrelet populations during the 1900s is accurate, how is it possible to tie this presumed decline to a decline in "old-growth?" particularly such "old-growth" conditions contained in the mature second growth stands of the Elliott? *In 1900 much of the current range of these animals consisted of hundreds of thousands of acres of burned snags and 10 to 50-year old even aged stands of Douglas-fir.* Animal grazing, agricultural fires, and federal fire suppression policies have resulted in many of the 1900-era stands continuing to exist to the present time. According to current popular wisdom, shouldn't the numbers of owls and murrelets and anadromous fish in these stands be measurably greater at this time?

In order to make effective long-term strategies for the maintenance of swimming and flying animal populations, it would seem that a regional consensus regarding the types and extent of such practices—and not just forestry practices— is needed. Even the most basic kinds of information is lacking: What is the record of these animals responses to catastrophic events in the past? Are these periodic "tests of rigor or viability" that these species need in order to retain their competitive advantage over other species? What role have introduced species played in this process? *What are the cumulative effects of human population growth, including industrial air pollution, urban lawn and park watering, highway construction, fencing, plowing, automobile use, and electrical power production on these animals?* How will riparian zone and tree age managment activities affect the over-all environment that these species exist in? The questions can't be answered by any one field of biology or any one forest plan.

Until we begin to better understand the complexity of what we're trying to do with our ESA strategies (apparently the goal is to stop the extinction process at virtually all costs), it would be safest—at least from an historical perspective—to not make sudden deviations from proven policies and activities: *These should be social decisions, not forestry decisions.* The costs of these decisions will be borne by millions of people not yet living: how will they view our (*society's,* not forestry's) decisions regarding these animals? Costly and quixotic? Or too little, too late?

E. Catastrophic fire history, forest productivity, and log income potentials.

It is important to remember that *the Elliott's contribution to coastal native wildlife habitat for the past 120 years has been a generally uniform, 80,000 acre, nearly even-aged, nearly pure, stand of young Douglas-fir trees.* This is a characteristic pattern for much of the western Coast Range, and one to which our animal populations have adapted over the past several thousand years.

The history of forest fires and reforestation in the Elliott portray an extremely productive and resilient environment for the growing of trees. The 1868 fire tells us, for instance, that *it should be possible to clearcut about 90% of the forest (over 80,000 acres) at one time—riparian zones and all—without an appreciable loss of productivity.* This claim is supported by the elimination of riparian vegetation thought to have occurred with the 1868 fire [ODF 1993: III-11], current timber production site class ratings [ODF, 1993: III-67], and the existence of current fish, owl, and murrelet populations. Such a clearcut, while unlikely to occur in a period of time less than several decades, might arguably result in measureable increases in native wildlife biodiversity, including birds, flowers, butterflies, and large mammals.

Because the primary goal of the Elliott Forest is to raise money for Oregon schools, it seems unnecessary that logging schedules be reduced so dramatically in a second growth forest at this time; a time when the state's public schools are reportedly suffering from Measure 5 cutbacks and our timber prices are at astronomically high levels (<u>over 500 times higher</u> than the first state timber sales of only forty years ago). At the least, it seems as if *"natural" conditions could be better obtained by integrating past clearcutting patterns into a better designed, more natural wind and fire-resistant pattern.* This type of planning could easily maintain the logging rates of the 1980s, take advantage of the sales prices of the 1990s, and rejuvenate the entire forest with a mosaic of wildlife habitat patterns that mimic the late 1500s and/or early 1700s.

F. Conclusions.

The outlined discussions on the previous pages were intended to provide a background to the more specific questions asked by OFIC. These questions are paraphrased below, followed by answers I believe can be supported by a current understanding of the Elliott's forest history.

1. How did the Oregon Department of Forestry use fire history for developing their recommended alternative?

ODF apparently did not use the 1770 to 1951 fire history described by Phillips in its development of plan alternatives, including Strategy 6. Rather, a 150-year *fire cycle model*—generally ascribed to Agee—seems to have been used to help justify a recent change in management focus from timber production to maintenance and creation of older forest conditions. If the available historical information had been used as the basis for regarding the Elliott's past and present, a much broader range of alternatives should have been developed for its possible and desired futures. ***A fire history perspective would have allowed for far greater latitude in the spatial and temporal designing of logging plans, wildlife habitat creation and maintenance strategies, and net income production.*** This conclusion is based upon a study of the general patterns and structural characteristics of forest wildlife habitat displayed over much of the western slope of the Oregon Coast Range during the past 300 years, by the demonstrated resiliency and productive capacity of the region's forest soils, and by the number and extent of its current wildlife populations.

2. How has the natural fire cycle shaped the Elliott State Forest?

The "natural fire cycle" of the Elliott supposes a mathematical predictive model that is biased against human activity (including logging, burning, and road building) and toward "average" decadent stand conditions, numerous older trees, and random lightning strikes. The principal problem with fire cycle models for planning uses is that they are based upon the faulty historical assumption that prehistoric families lived in an environment of big trees, uniformly-spaced snags, coarse woody debris, heavily shaded rivers and creeks, and randomly occuring forest fires. With fire cycle models, ecosystems function, snags are distributed, and random forest fires occur in an idealized, static, human-free environment in which homogenized mathematical patterns *(stages, averages, and cycles)* are used to represent past conditions, rather than the dynamic reality of documented fact *(history)*. The "shaping" of Elliott State Forest by fire can be better characterized *historically*, as *the result of a long-term series of botanical responses to constant human disturbances caused by daily and seasonal fires of varying degree and intensity.* The result has been an extensive stand of young, even-aged Douglas-fir trees that has persisted for over 120 years. Based upon a current understanding of the past three centuries of fire in the Elliott, it seems highly likely that a number of similar vegetation patterns have also existed for similar lengths of time in this same area during the past two to three thousand years.

3. Using an accepted fire history or fire cycle predictive model: What is the potential for shorter rotations to obtain the same wildlife objectives for spotted owls and marbled murrelets as given in the recommended alternative

If the "wildlife objectives" in Strategy 6 are interpreted to mean population maintenance or increase over time, *the mobility of owls and murrelets and their proven resiliency to catastrophic fires would indicate that historical rates of clearcutting can probably be maintained indefinitely* (but past fragmentation caused by this practice should probably be mitigated first); If the "wildlife objectives" are to simply create or maintain certain structural stand characteristics ("desired or required" habitat) within the forest, then these human definitions of desired future conditions would include (and require) a Strategy 6-type approach. Again, the difference in these two positions is the difference between *mathematical projections* based upon perceived *"averages"* and *interpretive projections* based upon documented *evidence* of disturbance and resilient recovery.

References

Bacon, Gerald 1987. Personal communications. Bacon is an older, life-long resident of Douglas County with an avid interest in local history.

Cronemiller, Lynn 1930. Annual Report of the Oregon State Forester. Oregon Department of Forestry, Salem, Oregon

Hines, Gustavus 1973. Oregon: Its History, Condition And Prospects. Arno Press, New York (facsimile reproduction of the original 1851 publication by Geo. H. Derby and Co., Buffalo, New York.)

Morris, William G. 1934 (December). "Forest Fires in Western Oregon and Western Washington," Oregon Historical Quarterly, Vol. XXXV, p.313.

ODF (Oregon Department of Forestry Coos District) 1993 (December). Elliott State Forest Management Plan Draft. Oregon Department of Forestry, Salem, Oregon.

Palmer, Joel 1983. Journal Of Travels Over The Rocky Mountains To The Mouth Of The Columbia River: Made During The Years 1845 And 1846. Fairfield, Washington.

Peterson, Jim 1993 (September/October). "In This Issue," Evergreen: p.2.

Phillips, Jerry 1988. Personal communications. Phillips is the most knowledgable individual available regarding the fire, forest, and logging history of the Elliott State Forest. Virtually all information on these topics in the Elliott plan draft was obtained from Phillips.

Pyne, Stephen J. 1982. Fire in America: A Cultural History of Wildland and Rural Fire. Princeton, New Jersey: 654 pp.

U.S. Coast and Geodetic Survey 1884. Report of the Superintendent of the U.S. Coast and Geodetic Survey, Showing the Progress Of The Work During The Fiscal Year Ending With June, 1883. Ex. Doc. No. 29, 46th Congress, Washington, D.C.

Zybach, Bob 1983. "Renewed Resources: The Reforestation Of The Tillamook Burn (1948-1983)," Associated Reforestation Contractors Quarterly, (Vol. 3, No. 4): 13-17.

———— 1988. The Great Fires Of The Oregon Coast Range (1770-1933) And Their Basic Effects Upon Current Vegetation And Property Ownership Patterns. Work in progress. Unpublished draft/outline copy on file at the Benton County Historical Society Library, Philomath, and with Fred Swanson, COPE Program, USDA Pacific Northwest Research Station, Corvallis. Original colored graphics and portions of later draft in possession of author: 237 pp.

VOICES In The Forest

Instead of finding an uninterrupted forest carrying 100,000 feet or more per acre reaching from the Cascades to the Pacific, the first settlers 75 years ago (1840) found in the valleys great areas of "prairie" land covered with grass, brakes or brush which were burned and kept treeless by the Indians, and mountain sides upon which forest fires had destroyed the mature forest and which were then covered by "second growth" of Douglas-fir saplings or poles.

Thornton Munger, forester, 1916, from "The Productive Capacity of The Douglas Fir Lands of Western Oregon and Washington," *The University Of California Journal of Agriculture*, Vol. 4, No. 3

The leading features of the Willamette Valley and the Tualatin plains were peculiar and strange to me as compared with any other country I had seen. Among the striking peculiarities was the entire absence of anything like brush or undergrowth in the forests of fir timber that had sprung up in the midst of the large plains, looking at a distance like green islands here and there dotting the vast expanse of vision. The plains covered with rich grasses and wild flowers looking like our vast cultivated fields, and where the rolling foothills approached the level valley these spurs would be sprinkled with low spreading oak trees, frequently with a seeming regularity that would seem unlike nature's doing, and at a distance like orchards of old apple trees.

James Neall, 1888, *A Down-Easter in the Far West: The Reminiscence of James Neall in Oregon and California,1845-50*, memoirs originally written to his niece, Abigail Rowell Tredick, printed in 1977 by the Oregon Book Society, Ashland

In larger burns, islands of timber along streams and in protected canyons often escape destruction. These supply seed to restock interior portions of the burns. Thus natural restocking following a single burn is the rule unless the fire be very large or severe. Each subsequent fire, however, greatly decreases the opportunity for natural reforestation. The interior islands of living trees, which escaped the first fire, are likely to be killed by the second, the existing seedlings are killed, and conditions made much less favorable for the establishment of seedlings from any seed that may reach the area.

Julius Kummell, Charles Rindt and Thornton Munger, 1944, *Forest Planting in the Douglas-Fir Region*, USDA, Forest Service

In the old days, there were vast areas where the Indian women burned in the morning of late summer and early fall. These areas were like gardens, kind of like the Garden of Eden that was described by the early pioneer people who came from Europe to the land...the landscape that people saw when they came from Europe was a landscape that was literally an expression of the culture of Indian people. When we talk about restoring the culture of the ten tribes I work with, we're talking about restoring the land as part of restoring the culture. There is absolutely no separation between the way the landscape looked in pre-contact times, the species composition and the structure of that forest, or that prairie, and the cultural needs and expressions of the Indian people. The land was an expression of the culture, as much as the arts and crafts and ceremonies.

Dennis Martinez, 1993, "Land and Culture", *Winds of Change: American Indian Education and Opportunity*, Vol. 8, No. 1

On the way, they met an old squaw, with a large firebrand in her hand, with which she had just set the grass and bushes on fire; when surprised, she stood motionless, and appeared to be heedless to any thing that was passing around here...there were no other Indians in sight.

Charles Wilkes, 1845, *Narrative of the United States Exploring Expedition during the years 1838, 1839, 1840, 1841, 1842*, Vol V

What would the forests of southwestern Oregon and northwestern California have looked like if no timber harvesting had ever taken place? We need only look at places where old growth has been undisturbed. We could assume that these remaining old growth stands are "natural," except that two significant historical factors are missing: indigenous people and periodic low-intensity fires. Both have been legally excluded from forests.

Dennis Martinez, 1993, "Land and Cuture"

This Countrey must be thickly inhabited by the many fiers we saw in the night and collums of smoak we would see in the day time and a delightful countrey thickly inhabited and Cloathed with woods and verdure with maney charming streems of water gushing from the vallies"

Robert Haswell, 1788, "Captain Robert Gray's First Visit To Oregon," *Oregon Historical Quarterly*

The woods are on fire below and above us (in Oregon City) and we are enveloped in smoke. We have had no rain during the last six weeks. Probably there will yet be two months without rain. There is some danger from the fire, it is so dry. Much timber is destroyed. The heat and dust are very trying during the day, but the nights are always cool and refreshing.

Reverend George Atkinson, from his diary, July 30, 1848

The forests of the north Pacific coast offer an exception to the law otherwise general, for this continent at least, that a change of forest crops follows a forest fire. The fir forests of western Washington territory and Oregon when destroyed by fire are quickly replaced by a vigorous growth of the same species, and the fires which have consumed great bodies of the California redwood have not prevented the reproduction of these species by sees and shoots.

Charles Sargent, 1880, *Report On The Forests Of North America, Exclusive of Mexico*, U.S. Department of Interior, Census Office, Washington, D.C.

In some cases, however, and especially in the largest burns, the work of reforestation has made little progress, owing probably to the difficulty of reseeding large burned areas. Since over many square miles all of the trees were killed, the seeds of a new crop have had to come from outside the region, and hence the seeding process has been slow. Areas are reported which were burned twenty-five and fifty years ago in where there is no vegetation larger than brush or ferns, trees of any species not yet having obtained a foothold.

Henry Gannett, 1902, *The Forests Of Oregon*, U.S. Geological Survey, Professional Paper No. 4, Series H, Forestry, 1; Washington, D.C.

Fires have widely ravaged the region examined. There is not a single forested township either on the west side or on the east side of the Cascade range in which the timber is not more or less fire marked. Without much doubt the present agricultural areas, once grass covered and carrying scattered stands of oak, were burned over quite as extensively as the timbered tracts; at least there are a few oaks that do not show fire marks. The only tracts that have escaped are the swampy sedge and tule-covered areas bordering the Klamath lakes and marsh, and such spots at the higher elevations where bare lava or pumice fields made the spread of fire impossible. Of the forest area examined, comprising in round numbers 3,000,000 acres, a total of 2,975,000 acres or 99.992 per cent are fire marked.

John Leiberg, 1900, "Cascade Range For est Reserve, Oregon, From Township 28 South To Township 37 South, Inclusive, Together With The Ashland Forest Reserve And Adjacent Forest Regions From Town ship 28, South To Township 41, South, In clusive, And From Range 2 West To Range 14 East, Willamette Meridian, Inclusive," *Twenty-First Annual Report Of The* United States Geological Survey, Part V, Forest Reserves, U.S. Department of the Interior, Washington, D.C.

Prairie soils, enriched by billions of decomposing grass roots, over thousands of years and protected from tree invasion by fire, grew the wheat and corn that fed America and a good part of the rest of the world. These soils also fed the European livestock that sustained the invaders. To deny this considerable contribution to the economic welfare of European peoples of North America is also to deny Indian people their place in the history of this continent.

Dennis Martinez, 1993, "Land and Culture"

We also took on board a considerable quantity of fine spars, for the Chinese market, where they are very much wanted and of course proportionately dear. Indeed the woods of this part of America are capable of supplying with these valuable materials all the navies of Europe.

Captain Meares, 1788, quoted from *General History Of Oregon*, Charles Carey, 1922

Reforestation is indispensable as insurance. Let us see to it that the untillable hills shall ever bear these matchless forests, emerald settings for our snow peaks. On their future depends, in great degree, the future of the Northwest.

John Williams, 1912, *The Guardians Of The Columbia*

At the time Cow Creek valley looked like a great wheat field. The Indians, according to their custom, had burned the grass during the summer, and early rains had caused a luxuriant crop of grass on which our immigrant cattle were fat by Christmas time...Fortunately in our case the land was ready for the plow. There was no grubbing to do. In all the low valleys of the Umpqua there was very little undergrowth, the annual fires set by the Indians preventing young growth of timber.

George Riddle, 1851, "History Of Early Days In Oregon," a series of articles by *The Riddle Enterprise*, 1920

Pollen analysis of...west central Oregon shows that postglacial forest succession differed from that in the Puget Sound region. This may have been due to the existence of forests in the Coast Range of Oregon during the latter part of the Pleistocene, and the occurrence of many periodic holocaustic fires during postglacial times.

Henry Hansen, 1941, "Paleoecology Of A Peat Bog In West Central Oregon," *Ameri can Journal Of Botany*, Vol. 28

At that time, there was not a bush or a tree to be seen on all those hills, for the Indians kept it burned over every spring, but when the whites came, they stopped the fires for it destroyed the grass, and then the young spruces sprang up and grew as we now see them.

Warren Vaughn, 1890, *Early Settlement Of Tillamook County*, unpublished memoirs by a Tillamook County settler, 1851-1867, typed manuscript provided by a staff member, Hebo Ranger District, Siuslaw National Forest

I think the largest single need in American forest biology is the study of man's relation to forest land. Our foresters need to understand much more than most of them do about purely human motives and aspirations with respect to the land. They ought to become genuinely knowledgeable and respectful of people's economic, social and aesthetic institutions.

Hugh Raup, quoted from Benjamin Stout, 1981, *Forests In The Here And Now: A Collection Of Writings Of Hugh Miller Raup, Bullard Professor Of Forestry, Emeritus, Harvard University*

Voices In The Forest: An interview with Bob Zybach

EVERGREEN: Mr. Zybach, your criticism of the Presidents's proposal for managing federal forests in the Pacific Northwest has caused quite a stir. Where did you find fault with the plan?
ZYBACH: There is nothing in the plan that suggests an understanding of this region's natural and human history. Because many of the plan's goals are founded on erroneous historic assumptions, the plan itself is fundamentally flawed.

EVERGREEN: Are you saying the Presidents's plan is doomed from the start?
ZYBACH: It is if its historical basis in not corrected. What is presented as history is both wrong and misleading. To make matters worse, there is a public perception this plan is based on the best available scientific information. Far from it.

EVERGREEN: What are the errors in history?
ZYBACH: There is a poor understanding of the role of fire in Douglas-fir forests. The extent of American Indian involvement in altering forest landscapes is downplayed, and the impacts of white settlement, particularly logging, are exaggerated.

EVERGREEN: And these historic errors adversely affect the President's plan?
ZYBACH: They certainly do. How can we expect to re-create pre-settlement forest conditions if we do not fully understand what conditions were present, or why those conditions were present?

EVERGREEN: What are "pre-settlement forest conditions"?
ZYBACH: Forest conditions thought to have existed before white settlement began in the early 1800's.

EVERGREEN: And the President's plan seeks to re-create pre-settlement conditions?
ZYBACH: That's correct.

EVERGREEN: Why?
ZYBACH: The plan strongly suggests that forests that were here before white settlement began were somehow more natural than are present-day forests. To accept this idea, you also have to believe pre-settlement forests were "naturally functioning ecosystems," untouched by human hands. This is their first big error in history. The fact is, people have been altering the character of this region's forests for at least 11,000 years.

EVERGREEN: How do these historic errors adversely affect the President's plan?
ZYBACH: The errors form the basis for a set of predictive assumptions about how natural ecosystems function, how old growth forests can be preserved, and how to maintain or increase plant and animal diversity in the region's forests.

EVERGREEN: What are these predictive assumptions?
ZYBACH: My review focuses on four predictive assumptions in the plan, but the main one assumes that when white settlement began, the Douglas-fir region was 60-to-70 percent covered with stands of old growth trees 200 or more years old. This is the so-called "sea of old growth" frequently discussed by proponents of old growth preservation.

EVERGREEN: How do you define "old growth?"

The Quillayute Prairie, in the Olympic Forest Reserve, Washington, 1899, from the 1900 U.S.G.S. report

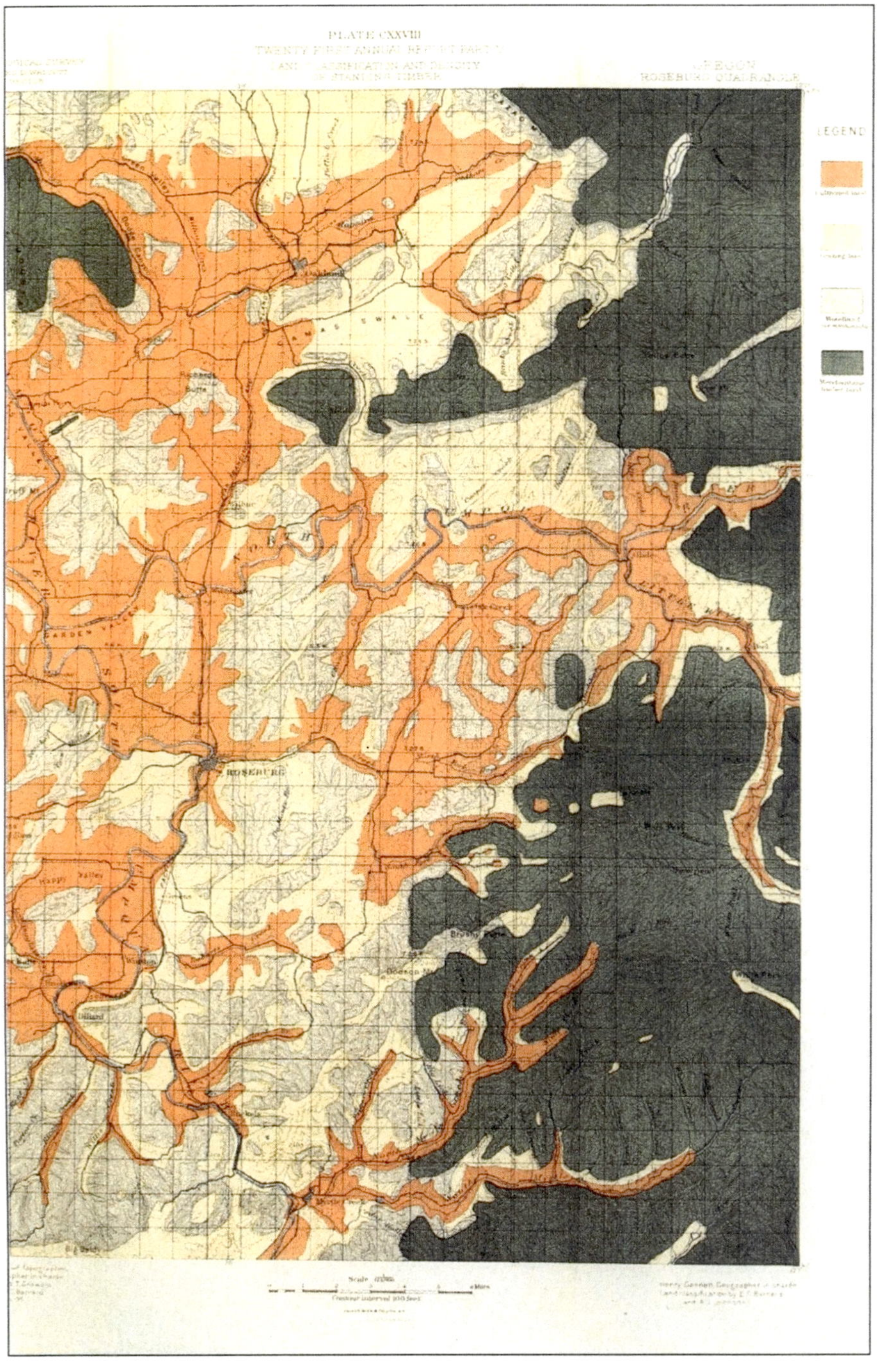

THE ROSEBURG QUADRANGLE, from the 1900 U.S. Geological Survey report. The area shown is about 26 miles wide and 76 miles long. Roseburg is to the left of center, just below the mid-point on the map. As the color-key indicates, a good deal of this quadrangle was cultivated land or grazing land. This is consistent with George Riddle's 1851 description of the Cow Creek Valley, which was reported in a 1920 story that appeared in The Riddle Enterprise. (See "Voices In The Forest") After studying this map and other historic references, Bob Zybach (See interview beginning on Page 7) said, "It is interesting to note that lowland stream corridors were not well shaded, yet salmon populations were thriving. This suggests that present-day opinions concerning the critical need for heavy tree cover in riparian zones may need rethinking. Sport and commercial fishing, and dam construction, have had far more to do with declining salmon populations than is generally acknowledged. Banning timber harvesting does not mean salmon will return. In 1900, salmon spawning streams passed through hundreds of miles of pasture land, where there were few trees, a result of Indian cultural fire, and cultivation by settlers."

ZYBACH: Old growth is a value judgement, not a scientific term. It means different things to different people. A hiker might see big old trees. A lumber grader probably sees high quality wood suitable for use as large timbers. A forest ecologist sees a range of timber stand characteristics. Others will see something else.

EVERGREEN: Where does age enter the old growth picture?
ZYBACH: That's a good question. The government's scientists started out seeing trees in the 180-to-220 year old range. Later they saw what they called "mature" structural characteristics they liked in much younger timber stands, even down to 80-year-old prime second growth.

EVERGREEN: Why the change?
ZYBACH: I don't know. It is never adequately explained in the plan. Whatever the reason, what started out as an attempt to protect old trees ends up appearing to be a colossal land grab.

EVERGREEN: How old are the trees in the old growth forest you picture in your mind?
ZYBACH: I am comfortable with trees in the 200-year-old range.

EVERGREEN: Earlier you said the President's plan includes four predictive assumptions that are wrong. What are the other three?
ZYBACH: The plan also assumes the American Indians living here lived in a "naturally functioning" environment. It is then assumed that logging is to blame for the presumed destruction of natural and cultural environments. Finally, it is assumed laws, forest ownerships and forest values in the Douglas-fir region are what they are today and will not change in the future.

EVERGREEN: What historic evidence can you cite that proves these assumptions are wrong?
ZYBACH: It can be shown that the so-called "sea of old growth" probably consisted of 200-plus-year-old trees growing in patches and stands that covered perhaps 5% to 38% percent of Douglas-fir region.

EVERGREEN: Based on the evidence you've uncovered, how would you describe the forests seen by early white settlers?
ZYBACH: Forests were more open than they are now. There were islands of even-aged conifers, bounded by prairies, savannas, groves of oak, meadows, ponds, thickets and berry patches. Many of these for-

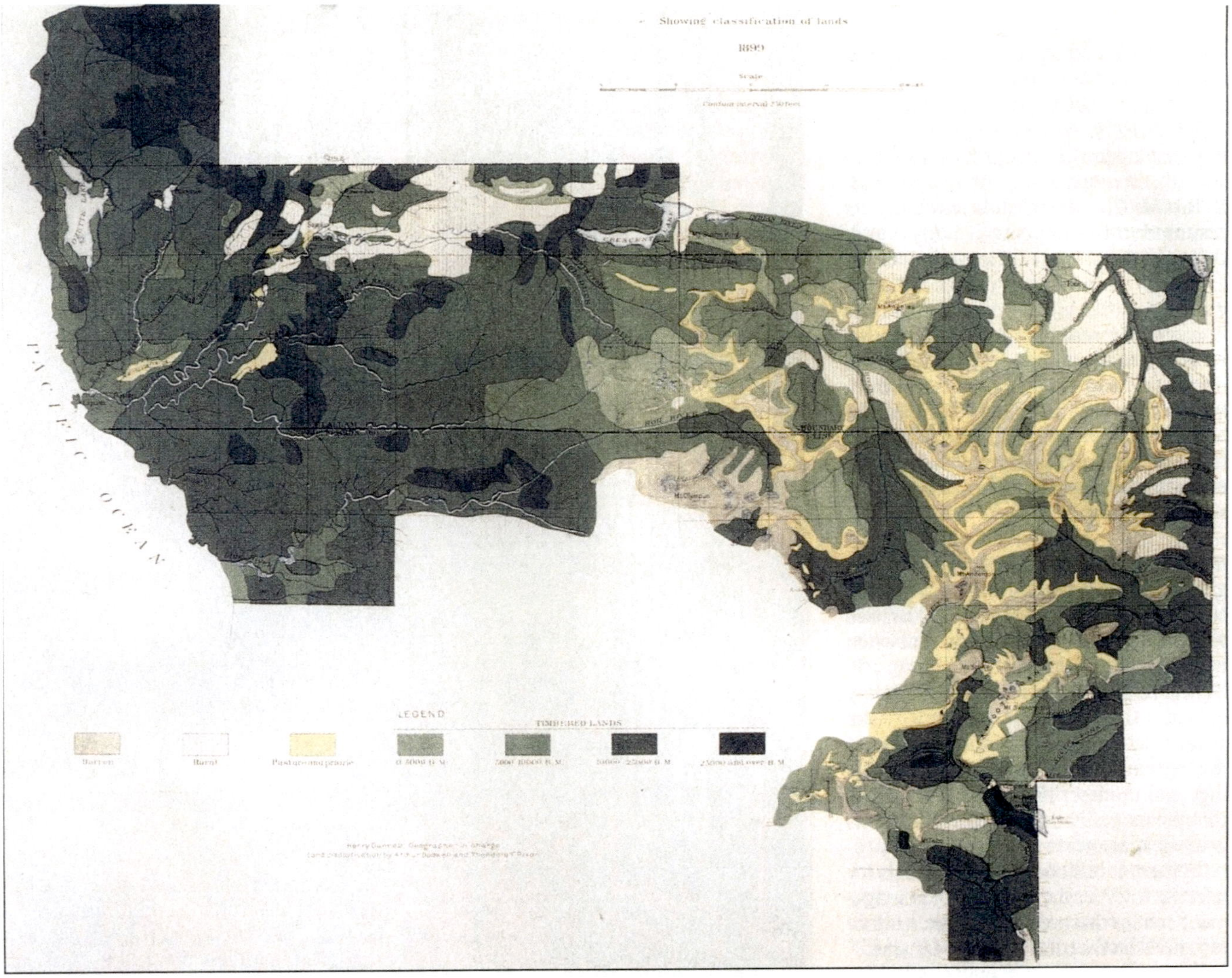

PART OF THE OLYMPIC FOREST RESERVE 1899, from the 1900 U.S. Geological Survey report. The area shown is about 60 miles wide and about 54 miles deep. Crescent Lake is in the top center, and the large open area to the left of the lake includes forest land that was burned in the 1907 Soleduck fire. The Soleduck River can be seen passing through the burn area.

The green areas below the 1907 burn have an impressive fire history dating back to 1667, and more than half of the Soleduck Ranger District has burned at least once since 1870. Maps showing the locations and approximate dates of these fires suggest this area could not possibly have been 60-70% covered with trees 200 years old, as the Clinton Administration forest plan indicates.

The yellow area on the far left is the Quillayute Prairie, shown in the 1899 USGS photograph on the facing page. Most of the lowland pastures shown on the map probably looked like the Quillayute Prairie. Salmon spawning streams passed through virtually all of these prairie pastures.

The darkest green areas on this map are the most heavily timbered. Lighter greens depict areas less heavily timbered. The 1900 USGS report includes detailed township-by-township descriptions, including tree species, volumes per acre and tree ages to the nearest year. The President's forest plan reports no such information exists for this period.

ests were virtually free of underbrush and coarse woody debris that has been commonplace in forests for most of this century.

EVERGREEN: Your description is a far cry from what is contained in the President's plan.

ZYBACH: This is not my description. It is what is revealed in a rigorous review of this region's human and natural history, and yes, it is much different from what is described in the President's plan.

EVERGREEN: You mentioned Indians a few moments ago. The President's plan doesn't say much about Indians who lived here before white settlement began. What does the historic record show?

ZYBACH: The plan is wrong in its contention this region's forests were largely untouched by human hands before white settlement began. Humans have lived here as families for at least 11,000 years. The use of fire by these families produced a forest environment dominated by fire-dependent and fire-tolerant plant species.

EVERGREEN: But didn't Indians live in harmony with nature?

ZYBACH: That's a very subjective question. It assumes some sort of delicate, undisturbed balance in nature. I don't buy it. This region's forests are not the product of some gentle and unseen hand moving gracefully across landscapes. These forests are the product of vast and frequently violent natural disturbances, including human-caused fires of almost unimaginable scale.

EVERGREEN: Let's go back to the Indians. Didn't they live in harmony with nature?

ZYBACH: I guess that depends on how you define harmony. Indians were very aggressive resource managers, just like all civilizations and all peoples. There is widespread evidence of their impacts on this region's landscape. From the Puget lowlands south into northern California, they may have burned more than a million acres a year. Indians used fire to create habitat for wildlife, to clear away trees and underbrush, and to shape forests in the image of their own culture.

EVERGREEN: What about the assumption that logging is responsible for the presumed destruction of forest ecosystems?
ZYBACH: The plan is quite specific in its assumption that logging is the primary culprit in the presumed destruction of forests, and again history proves the plan wrong. A more objective look at logging's impacts on forests suggests that logging, broadcast burning and tree planting can mimic prehistoric patterns of forest fires, landslides, windstorms and volcanic eruptions. This means we should be able to use these human interventions to help recreate natural conditions present when white settlement began.

EVERGREEN: Are you suggesting logging and broadcast burning and aggressive reforestation programs could be used to recreate forest conditions present when white settlement began?
ZYBACH: The historic record certainly indicates this is possible, but because the government's scientists missed most of this region's human and natural history, they end up describing forest conditions that never existed. They then use this grossly inaccurate description of pre-settlement forest conditions as a basis for offering the President a series of management options that are not more than minor variations on the same three-part theme.

EVERGREEN: And what is this three-part theme?
ZYBACH: The plan's authors would have you embrace their assumption that the best way to manage old growth is to preserve it in refuges. Related to this is the assumption that the best way to protect native wildlife diversity is to preserve old growth. Related to this assumption is the assumption the best way to delay or stop the extinction process is to delay or stop logging old growth.

EVERGREEN: And you disagree with these themes?
ZYBACH: Whether I agree or disagree with these themes is not the question. The fact is a rigorous review of this region's natural and human history will show the government's scientists are wrong. Their plan will do little to protect old growth forests, and may place forest ecosystems at greater risk than would a seemingly more aggressive strategy.

EVERGREEN: So you are in agreement with Dr. Chad Oliver's belief that a wider range of harvesting and thinning strategies could actually speed the creation of structural diversity in forests?
ZYBACH: I am not academically qualified to evaluate Dr. Oliver's work, but I can tell you that the historic record is very clear, and the historic record agrees with what Dr. Oliver is suggesting.

A stand of fir and hemlock on the Soleduck River, three miles above Hot Springs, from the 1900 USGS report. The stand was leveled in the 1907 fire, and here it shows evidence of previous burns. Note the lack of vegetation on the forest floor.

EVERGREEN: Mr. Zybach, is there an abundance of historic evidence that supports your criticism of the President's plan?
ZYBACH: You can literally measure the evidence by the pound, and I have collected several thousand pounds that I am now using in my research. The evidence divides itself nicely into four periods: The prehistoric period, beginning 12,800 years ago, and continuing to 1774; the early historic period, running from 1778 to 1859, a more recent historic period, which ran from 1860 to 1945; and the modern era, which began after World War II.

EVERGREEN: Let's take these periods one at a time. What evidence can you cite?
ZYBACH: The prehistoric record is revealed in pollen fossils, archaeological findings, scientific evidence of early cultures present in the region, tree ring analysis, and analysis of plant populations now living in environments they did not inhabit in the past. Here I should note that many of the scientific techniques used for analyzing this sort of information have been in use for more than 50 years.

EVERGREEN: Can you give us a specific example of evidence from this prehistoric period?
ZYBACK: Yes, I can. Jan Henderson, who is a plant ecologist with the Forest Service, wrote a paper in 1990 titled *Trends In Amount Of Old Growth Forest For The Last 1000 Years In Western Oregon And Washington*. His analysis of fire history records for the Mt. Baker-Snoqualamie

and Olympic national forests reveals a relatively stable 1,000 year trend in the amount of old growth. And his estimate for the year 2,000 is 40% old growth, a number that exceeds the average of his projections for the same areas between 1300 and 1800 A.D. Moreover, his estimates for old growth cover for three western Washington and five western Oregon national forests show increases in post-settlement old growth over the last century.

EVERGREEN: What do you conclude from Mr. Henderson's research?
ZYBACH: Two conclusions are possible. First, there is more old growth now in some parts of the region than there was before white settlement began. Second, some things happened around 1650 that precipitated a significant amount of natural regeneration, as well as the generation of forests where none had been present before.
EVERGREEN: What do you think happened?
ZYBACH: There is widespread evidence of catastrophic wildfire, both in western Washington and in western Oregon. Some of these could have been million-plus-acre fires, and the on-the-ground evidence uncovered thus far indicates these were very hot fires, a far cry from the gentle under burns hypothesized in the President's plan. More field work is needed before we can fully grasp the role these fires played in shaping this region's forests. In the absence of this information, it is not possible to make accurate predictions about future forest conditions, or what management strategies are needed to protect old growth forests.

The photo, from the 1900 USGS report, was labeled, "Severe burn on mountain side." The location was not given, but the fire-killed snags appear to be quite small, and the stand quite dense, suggesting repeated burning.

EVERGREEN: What sources are available that describe more recent forest conditions?
ZYBACH: Well, we go next to the period beginning in 1778 and ending in 1859. This is an important period because it includes the so called "white settlement" and "pre-logging" periods specifically used to model the ten preservation alternatives presented to President Clinton. Here we find the "native forests" and "naturally functioning ecosystems" that are evaluated for each alternative as a basis for what the government's scientists called "an expected likelihood of achieving long term past conditions."

EVERGREEN: And what source material have you found for this period?
ZYBACH: In addition to the same sources that exist for the prehistoric period, we have qualitative descriptions written by some of the region's earliest explorers, including Robert Haswell, Lewis and Clark, Alexander Henry, David Douglas and the Wilkes Expedition. Their first-hand accounts of forests and peoples living here then present a much different picture than is presented in the President's plan.

EVERGREEN: Most of these people were not forest scientists. They were explorers.
ZYBACH: That's true, but their descriptive accounts match up quite nicely with our earliest scientific surveys of the region. We also have maps and notes from General Land Office surveys that began in 1851. Tree species were identified, measured for diameter and mapped at half-mile intervals. There are also descriptions of what grew beneath the trees, human developments, crops and general landscape features.

Perhaps the most interesting of all the old scientific sources I found is a U.S. Geological Service book titled *Annual Reports of the Department of the Interior, for the Fiscal Year Ended June 30, 1900.* It includes detailed forest inventory data, plus a beautiful set of full-color maps showing the distribution of various types of vegetation then growing in the region. These old

A portion of the 1914 Oregon State forest type map

In 1914, Oregon State Forester, F.A. Elliott, allocated almost $7,000 - an amount greater than his annual salary and travel budget - to the "compilation and printing" of a state map.
Thirty-five years later, in 1954, John McWade a night dispatcher with the state forestry department, rescued the map from a trash bin. It survives today, and provides a wealth of information about forest condition, statewide, at the turn of the century. Among the map's many details: the location and extent of prehistoric and historic fires, logging boundaries, commercial timber stands, successfully reforested areas, early roads, trails, telephone lines, towns and legal survey boundaries.

Like other historic records, the 1914 map depicts a much different forest than the sea of old growth timber repeatedly described in the Clinton forest plan.

Nearby, are color photographs of portions of the 1914 map, plus a digitized version of the map. The digitized map is somewhat confusing because too much information has been overlaid on a single display. For example, the green area designated "merchantable timber" includes all timber of varying age and species, including young second growth saplings and old growth timber. Areas that appear darker green are federally protected as habitat or wilderness.

Although this map is difficult to decipher, it does illustrate the power of computers - in this case geographic information system (GIS) technology - to integrate present-day scientific data, including high altitude satellite imagery - with earlier historic records, including photographs, maps and on-the-ground timber cruises done almost 100 years ago. Integration of past and current data could help land managers forecast a possible range of desired future conditions.

The area shown in the digitized map is bordered on the north by the Columbia River, on the south by the Oregon-California state line, on the east by State Highway 97 and on the west by the Pacific Ocean. The Clinton forest plan designates everything west of the solid red north-south line as spotted owl range, and everything west of the dotted red north-south line as primary marbled murrelet range.

The digitized map was prepared for illustration purposes by Richard Crucchiola, senior geographic information systems programmer for the Oregon State Service Center. Additional historic and technical data was supplied by Bob Zybach.

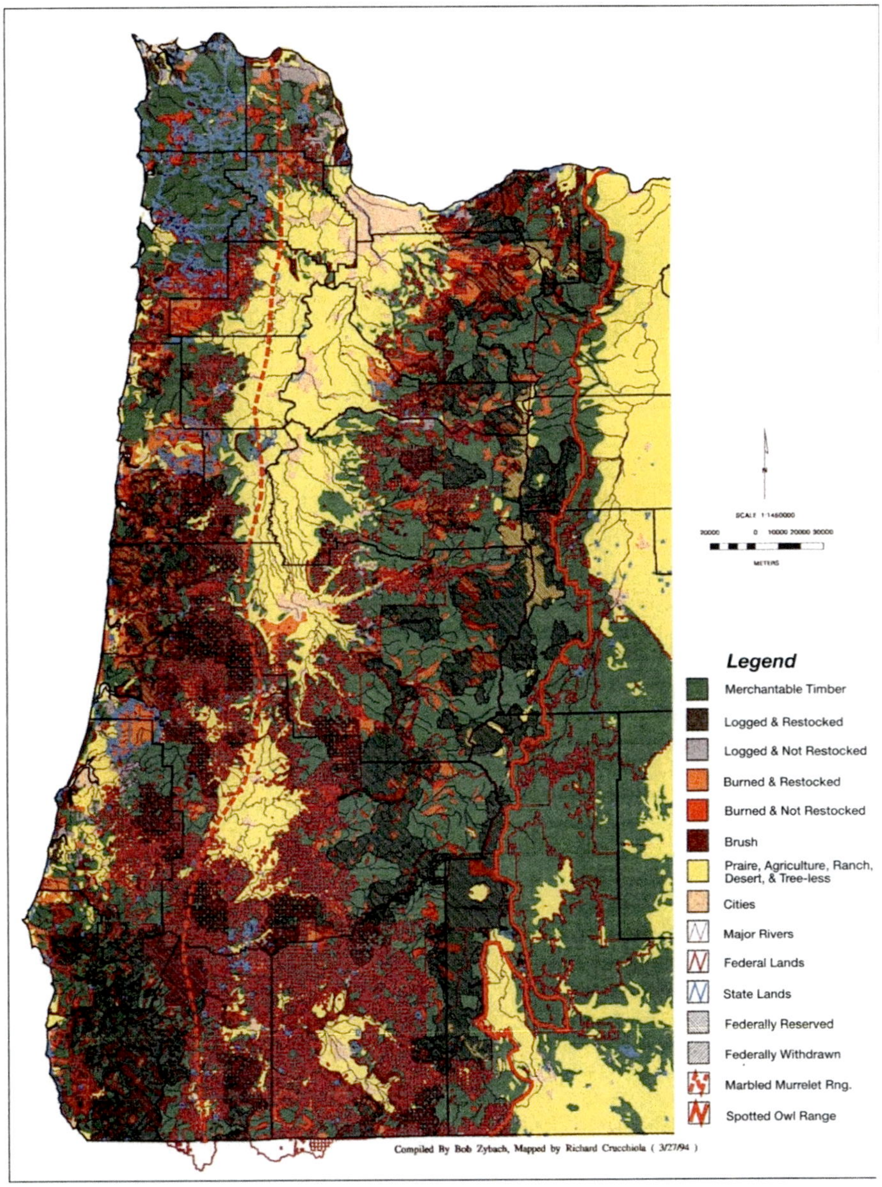

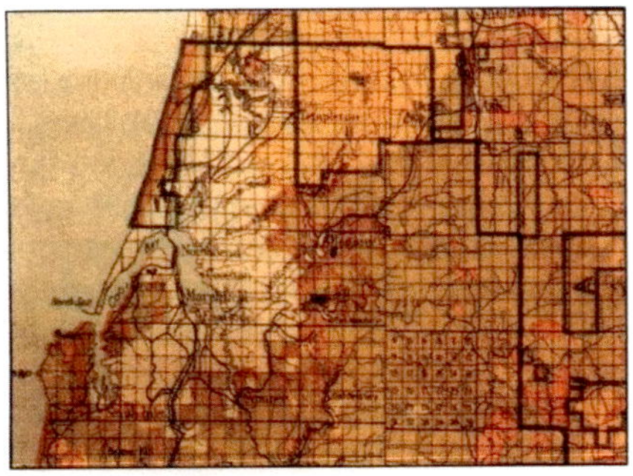

reports describe a much different forest than is described in the President's plan, and according to the plan, none of this data exists.

EVERGREEN: How could the government's scientists have missed this information, and how did you find it?
ZYBACH: I can't speak for the government's scientists. I have a library card.

EVERGREEN: Are there other recent sources that describe a different history than that described in the President's proposal?
ZYBACH: Yes, there are other sources, especially for the all-important 1860-1945 period. Here the evidence includes first-hand accounts written by the region's earliest professional foresters. These accounts were ignored or overlooked, as were literally thousands of historic photographs, early aerial photographs, tax records, land survey records and living memory accounts from diaries and newspaper stories written by people who helped settle this region.

EVERGREEN: Why do you say this period is all-important?
ZYBACH: Because it connects our earliest conifer plantations with an earlier time when forests reseeded themselves naturally. Also because historic records from this period could be used to document changes in wildlife habitat patterns and timber volume densities. Timber stands established during this period were all heavily modified by human actions, including a 1910 Congressional decision to start putting out forest fires, the demise of Indian cultures, farm clearings, fencing, grazing, excessive hunting, clearcutting and our earliest reforestation efforts.

These timber stands are our forests now, and many of them are being groomed to become the next generation of old growth, yet the government's scientists would have you believe these forests are the remnants of "naturally functioning, pre-settlement ecosystems." They aren't. They are creations of human actions and human interactions with nature.

EVERGREEN: So these aren't old growth forests?
ZYBACH: Again, it depends on how you define old growth. Some of these forests are truly old, and do meet the structural characteristics cited in one or more of the generally accepted definitions of old growth forests. But there are also prime second growth forests, and until the spotted owl came along, it had been assumed these forests would be managed forests,

Log jam in Montana's Blackfoot River, from the 1900 USGS report. This is the river the late Norman Maclean wrote about in *A River Runs Through It.* He learned to fly fish in the Blackfoot and considered it to be one of the finest trout rivers in Montana, a state noted for fine trout waters. He fished here frequently throughout his lifetime. The presence of these logs in a fine trout river - where they should not have been - bears witness to the resiliency of nature.

available for harvesting in perpetuity. Now we learn this is not the case.

EVERGREEN: What is the most important point you are trying to make here?
ZYBACH: That virtually all of the forests we see today in the Pacific Northwest are the product of human intervention or human interaction with nature.

EVERGREEN: But what about all of the one and two and three-log loads we've seen running up and down highways in the Pacific Northwest for the past 30 or 40 years? Surely these trees predate white settlement?
ZYBACH: Some of them probably do, but that does not prove the region was once a vast sea of old growth. An easier way to gain a perspective on the ages of forests is to simply count backwards.

EVERGREEN: For example?
ZYBACH: For example, forests that were 200 years old in 1950 were only 50 years old when white settlement began. It would be a real stretch to say a 50-year-old forest growing at the time white settlement began was old growth. Therefore, its subsequent development into commercial old growth timber may well have been a function of settlement, rather than any other naturally occurring event. A region-wide inventory of tree rings on the stumps of trees cut during the past 50 years is needed to quantify the cumulative impacts of logging and fire suppression in old growth forests. School children could do this work. So could unemployed loggers.

EVERGREEN: Earlier you mentioned a set of full color maps that were part of a 1900 U.S. Geological Survey report. What do these maps reveal?
ZYBACK: The maps indicate that in the early 1800's, when white settlement began in the region, significant portions of what is now spotted owl habitat contained only small amounts of trees 200 years old or older, perhaps no more than five or ten percent. These maps also indicate southwest Oregon was much more open than it is presently. Pioneer diaries support this fact, and it is known that early settlers

chose the valleys that parallel the Interstate 5 corridor because these valleys were open and could be tilled with a modest amount of clearing.

EVERGREEN: How do you explain all of the old postcard photographs showing entire logging crews standing on single stumps large enough to hold a dozen men?
ZYBACH: There are still trees this big in some areas, particularly along the coast where growing conditions are ideal. But these trees were exceptions, not the norm. It often took days to fall a single tree of the size you describe. Early day sawmill records indicate most trees sawn were from 75 to 200 years old, depending on the species. These trees were not the huge old trees you see in photographs. Also, it is important to remember size can be deceiving. I have a photograph of a felled tree that is five and one-half feet in diameter at its base. It was 112 years old. I know because I counted the rings.

EVERGREEN: How much old growth do you think there was in the Pacific Northwest at the time white settlement began?
ZYBACH: In my critique of the President's plan, I estimated that from 5% to 38% of the region contained trees 200 or more years old. Based on my own research, I would also estimate the region's forest "blanket" is 10% to 20% larger than it was 150 years ago, with most of the increase due to the growth of forests in areas once occupied by old burns, prairies, savannas and meadows.

EVERGREEN: Can we assume there is more old growth in the region's forests now than there was before white settlement began?
ZYBACH: Not really. Historic records can be used to support a view that there is less old growth or a view there is more old growth. It depends on which old growth definition is used and which tree species are considered. What is true is that there will be more old growth in the future as more so-called mature timber stands begin to take on the definitional characteristics of old growth.

EVERGREEN: Let's go back to your earlier statement about forests invading areas that were once prairies and savannas. How did this happen?
ZYBACK: The prairies and savannas were largely a product of Indian cultural fires, which ended about the time white settlement began. Later, farm tractors replaced draft animals, and horses gave way to the automobile. Pasture land was abandoned and in the absence of fire, trees invaded the old pastures. The migration from rural towns to cities began, and we started putting out forest fires. The combination of these events allowed forests to spread over once open landscapes. These same events also allowed forests to grow more dense, with much larger accumulations of underbrush and dead woody debris. This

Washington log and loggers, circa 1925, a postcard sent from Kelso, Washington by Paul Petersen, a millwright, to his young son, Darrell. Back then, big logs like this were usually burned in steam donkeys because saw mills were not yet able to handle them. Smaller logs were preferred because they were more easily turned into cord wood, railroad ties and bridge timbers. (Jim Petersen collection)

The caption for this photograph from the 1900 USGS report reads, "Common method of hauling yellow-pine logs to sawmill." Compare the size of these logs with the log and loggers shown in the nearby 1922 postcard. Early sawmill records indicate logs of the size shown on the postcard were the exception rather than the rule. Most of the earliest logs were turned into railroad ties, mine timbers, firewood and bridge pilings, products best made from smaller trees. Besides, the equipment needed to transport the largest logs to mills did not yet exist, and they were frequently burned as fuel in "steam donkeys," steam powered machines used to drag logs from hillsides to railside.

is why the risk of catastrophic fire is becoming so great in the region's forests.

EVERGREEN: This question is unrelated to your critique of the President's plan, but what do you think of the idea of re-introducing fire in these forests to clear out the deadwood?
ZYBACH: I suppose it is possible in some places, and indeed this is talked about in the President's plan, but we have millions of acres where woody debris accumulations are simply too great to be burned safely. Tree thinning would be less risky. Also, my guess is the fires the government's scientists want to set will be judged violations of the Clean Air Act.

EVERGREEN: Let's go back to errors in history. How do we square your historic perspective with the oft-repeated claim that harvesting has so fragmented the region's forests that wildlife has become isolated on islands of trees that are all that remains of what was once a sea of old growth?
ZYBACH: You can't square this claim with what history records because the sea of old growth never existed. Several early pioneers wrote vividly about the "green islands" of trees they saw that dotted vast savannas in the present day Tualatin and Willamette valleys. The still unanswered question is, "Can islands of trees created by logging provide the same type of wildlife habitat as islands of trees created by wildfire and Indian fire in the pre-settlement era?"

EVERGREEN: What does your research reveal about the impact of Indian fire on pre-settlement forests?
ZYBACH: I'll get right to the point. If the government's scientists were to construct an accurate portrayal of Indian land management practices, their sea of old growth myth would vanish in a puff of smoke. The fact is Indians regularly burned the landscape to create garden-like settings, which were quite clearly described by early white settlers. Indians used fire to create and maintain vast wildlife habitats, yet these habitats are not even being considered in today's ecosystem management discussion. Indian fires also minimized the amount of dead woody debris present in pre-settlement times, reducing the risk of catastrophic fire.

Denying that this happened, on the scale that it happened, is denying the obvious. Until Indian impacts on this region's forests are fully discussed and understood, there is no way the government's scientists can mimic, restore or maintain past forest conditions they write about in such glowing terms.

EVERGREEN: Is there much in the way of specific information about actual forest conditions that existed before white settlement began?
ZYBACK: Yes there is, and if you study the evidence as I have, the idea that we should be managing forests along more natural geographic lines, rather than property lines, begins to make sense.

EVERGREEN: What do you mean?
ZYBACH: I mean that natural forest conditions appear to define themselves, and their differences, on a landscape basis, and for purposes of definition, I define a landscape as a major river drainage.

EVERGREEN: So you are saying that differences in forest landscapes are most noticeable as you go from one major river drainage to the next?
ZYBACH: That is certainly the case in western Oregon and western Washington. One major reason these drainages defined differences in forests is that they frequently formed major barriers to human-caused fires. Even big fires burn best uphill, and when they get to the tops of mountains that border major drainages, they generally lose their momentum. This pattern helps explain why we usually find pockets of truly old trees in the bottoms of canyons where they were protected from high winds and big fires.

EVERGREEN: But what specific information have you found that gives us a glimpse of forests that were here years ago?
ZYBACH: In 1914, Oregon state forester, F. A. Elliott, commissioned development of a map of the state. The map still exists, and it provides a one-of-a-kind view of Oregon's forests at the dawn of the age of the automobile. The extent of prehistoric and historic forest fires is shown, as are logging boundaries, commercial timber stands, successfully reforested areas, early roads, trails, telephone lines, towns and legal surveys. It is a unique view of pre-World War I Oregon, as seen from the air.

EVERGREEN: What can we learn from this map?
ZYBACH: We can learn a great deal about the type and location of fire-dependent and fire-tolerant plant species, and we can learn a good deal about where harvesting had occurred, where forests have been replanted, where merchantable timber was located, which lands had not been replanted after harvest, and where brush was dominant. In short, what we have is a detailed vegetation type map. Things get even more interesting when you compare this map to earlier historic accounts describing forests.

EVERGREEN: And you have done this?
ZYBACH: Yes, I have, and our earliest historic writings correlate nicely with what the 1914 map shows.

EVERGREEN: Tell us more about these early historic writings.
ZYBACH: We have tree measurements taken along the Columbia River in northwest Oregon by Lewis and Clark in 1805 and 1806, David Douglas in 1825 and 1826, and by the Wilkes expedition in 1841. We know Indians burned less frequently in this area, and the result was a towering forest of Douglas-fir, cedar and hemlock. Those who lived here were more dependent on fish. Today, corporate tree farmers own most of this forest land.

We know also that the Oregon Coast Range, extending south to the middle fork of the Coquille River, included vast stands of even-age, nearly pure Douglas-fir. This area was the scene of some of the largest and hottest fires in history and produced some of the fastest growing conifers ever measured. Indian cultural fires had a profound influence here. One of our earliest accounts describing Indian fires is from a daily journal written in 1788 by Robert Haswell, an officer on the *Washington*, Captain Robert Gray's fur trading sloop. Today, this landscape includes the Elliot and Tillamook state forests, BLM-managed O&C lands, and Siuslaw National Forest.

We also have accounts describing the interior valleys, including the Sacramento, Rogue, Umpqua, Willamette and Puget. Here Indians burned hundreds of thousands of acres annually, and the result was a nearly contiguous series of great prairies and oak savannas extending almost the entire length of the Cascade Mountains. Our earliest descriptions of these areas were written in about 1826. Today, ownership here is concentrated in agriculture, urban development, private second growth tree farms and O&C lands.

The same kind of information exists for all of Oregon and Washington, including the national forests now identified as critical habitat for the northern spotted owl.

EVERGREEN: Are there any references to the 1914 map in the President's plan?
ZYBACH: No, there aren't.

EVERGREEN: How about these other documents you've talked about, the old forest inventory records, the diaries, the photographs, anything?
ZYBACH: None of this is mentioned in the President's plan. It is no wonder so-called pre-settlement forest conditions described by the government's scientists are much different than what is described in actual history. They didn't do their homework.

EVERGREEN: What kind of forest do these scientists describe?
ZYBACH: A sea of old growth, functioning naturally in perfect harmony, always moving toward a state of equilibrium. Scientists began to discard this notion 50 years ago, and to replace it with a belief that nature exists in a state of constant change, if not constant turmoil. This has certainly been the case here in the Pacific Northwest, where fire has been the dominant natural force in shaping forest landscapes.

EVERGREEN: Have there been other kinds of natural or non-human disturbance that have helped shape the region's forests?
ZYBACH: There certainly have been. Catastrophic floods, hurricane-force winds, widespread insect outbreaks, volcanic eruptions, massive landslides, beaver ponding and climate changes have all influenced the character of our forests, and thus the location and sizes of various wildlife populations.

EVERGREEN: Are the effects of these disturbances discussed in the President's plan?
ZYBACH: There is barely a mention.

EVERGREEN: Based on your research, what can you tell us about these disturbances?
ZYBACH: Geologic records show that the Willamette Valley filled with water perhaps 50 to 100 times between 15,000 and 12,800 years ago; and we know that about 12,800 years ago, one of the greatest floods in world history roared down the Columbia from melting ice fields in Montana. Soil deposits left by floods account for the flattened topography of today's Willamette Valley.

Pollen counts, taken from trees that died thousands of year ago, reveal a history of changing landscape patterns in western forests. Great winds also significantly altered the landscape, as did big fires. These winds include the 1962 Columbus Day storm, the "Big Blow" of 1921, and other major windstorms in 1788, 1880, 1895, 1923, 1961, 1979 and 1981.

Snow was also a big factor. The snows

Summit of the Siskiyou Mountains, near Sterling Peak, from the 1900 USGS report. In memoirs written to his niece in 1888, Oregon pioneer, James Neall, described, "...the entire absence of anything like brush or undergrowth in the forests of fir timber that had sprung up in the midst of large plains, looking at a distance like green islands here and there dotting the vast expanse of vision." What he was describing were scenes like this one - forests amid prairies and savannas, created by Indians who burned the land regularly to prevent forests from spreading into open areas frequented by deer and elk, which the Indians hunted.

of 1861, 1881, 1882, 1936 and 1937 wiped out huge livestock herds grazing in the Willamette Valley. With fewer animals to keep the prairies clear of underbrush, Douglas-fir spread into hillside pastures.

We also have evidence of long periods of drought, beginning in 1717. Sustained drought patterns correlate nicely with major forest fires that burned throughout western Oregon.

If you add these up - the winds, snows, droughts and floods - you have a substantial amount of natural disturbance within what is now the range of the northern spotted owl. But for reasons that are never explained, the President's plan ignores these natural events, and their impact on forest habitats present in forests today.

EVERGREEN: How do you think fires occurred in pre-settlement times?
ZYBACH: I am not a proponent of the idea that fires came and went in cycles. Keep in mind that cultural fire was a daily occurrence in this region for thousands of years. Indians cooked on these fires, and they warmed themselves with fire. They also burned seasonally, in the spring and the fall, to clear away trees and underbrush and to stimulate the growth of wildlife forage. What this means is that there were constant sources of ignition. It is reasonable to assume many of the catastrophic forest fires for which we find evidence were probably set by Indians intent on clearing land, controlling the spread of Douglas-fir, and creating habitat for wildlife.

EVERGREEN: What does this record of natural and human-caused catastrophe tell you about spotted owls?
ZYBACH: I'm no wildlife biologist, but common sense tells you owls are survivors, able to adapt to a wide range of constantly changing forest conditions. I recently read that some of the highest reproductive rates thus far recorded for owls were on private industrial tree farms.

EVERGREEN: Is there evidence humans have been affecting wildlife habitats for a long time?
ZYBACH: There certainly is, but with the exception of logging, the government's scientists virtually ignored the impacts of 11,000 years of human influence on wildlife. Yet these disturbances clearly contributed to the extirpation or extinction of many wildlife species, including grizzly bears, camass, lampreys, deer, elephants, camels, giant sloths, and giant beavers that were hunted by Paleoindian hunters during the last 11,000 years. Human disturbance also contributed to the introduction of other species now considered to be a part of the wildlife population, including bullfrogs, elk, bachelor buttons, orchard grass, opossums, many introduced tree species, and even rainbow trout.

Amazingly, the only human disturbance discussed in the plan is logging, and what is written reveals a deep-seated anti-logging bias and a poor understanding of natural history.

EVERGREEN: Explain what you mean.
ZYBACH: Well, you can start with the President and Vice President, and the prepared statements they read at the timber summit. Both men talked about "precious old growth forests" that "once destroyed can never be replaced" and are "gone forever." Whose information are these views based upon and why have they been so completely accepted? There is little evidence in human or natural history to support these viewpoints. This rhetoric is based on very selective use of information that supports a romantic vision about what pre-settlement forests looked like.

EVERGREEN: Are you suggesting logging has had no impact on this region's forests?
ZYBACH: No. Flat out, no. Logging has had a profound influence on the structure and extent of the region's forests.

EVERGREEN: Give us some examples.
ZYBACH: Commercial logging began in the Douglas-fir region at least two generations before white settlement began, probably in 1788 on Vancouver Island. Then, in 1810, a fur trading company cleared trees in the Astoria area; and in 1827, the Hudsons Bay Company built the first commercial sawmill west of the Mississippi near Fort Vancouver. None of this history is mentioned by the government's scientists. Instead, we are told white settlers burned forests or hacked them down to clear land for farming. This is simply not true. Early white settlers regularly avoided forests in favor of prairies because the prairies were ready for plowing. These prairies were, of course, a product of the Indian cultural fires the plan minimizes.

EVERGREEN: So what is your point?
ZYBACH: My point is simply this. Without a clear idea about why logging took place, where it took place and when it occurred, it is impossible to understand or measure its long term effects. By completely ignoring the beneficial aspects, and by misrepresenting the role of logging in northwest history and culture, the government's scientists set the stage for a plan that is biased against logging.

EVERGREEN: What benefits do you believe logging has produced in the region's forests?

ZYBACH: Again, you need to understand the great difference between forests described in the President's plan and forests described by people who actually saw or lived in these forests.

Fire-killed alpine hemlock, noble fir and lodgepole pine, near the headwaters of the North Fork of the Rogue River, from the 1900 USGS report. Not much in this picture hints of giant trees or a vast sea of old growth timber. There were, and still are, some very large trees in this area. They got big by surviving repeated fires, caused by lightning or set by Indians.

Yellow pine stand along the east side of the Williamson River in the upper Klamath River Basin, central Oregon, from the 1900 USGS report. Note the lack of brush beneath the trees. Fire kept the ground clean and the grass growing.

The plan strongly suggests clearcutting creates even-aged stands of young trees that represent an alien and impoverished environment for wildlife populations. In a report he helped write in 1988, Dr. Jerry Franklin, who led the government's scientific team, wrote, "In the early part of this century, most of the forested area west of the crest of the Cascade Range was covered by old-growth forests consisting of Douglas-fir, western hemlock...and several other large, long-lived conifer species. Most of these forests were probably more than 300 years old and many exceeded 750 years."

Dr. Franklin's repeated use of the word "most" suggests to me that he believes more than half the region's forest land base was covered by old growth at the turn of this century, and that most of this turn-of-the-century forest was about 400 years old. He is not even close. There is not a chance in the world the forest he describes has existed here in the past 10,000 years.

EVERGREEN: That's pretty tough talk for a graduate student.
ZYBACH: Again, the historic record speaks for itself. Thornton Munger, who was the first renowned forest scientist to live and work in this region, described a much different forest in a report he wrote in 1940. Listen to what he said. "The paths of the great forest fires of the last century or two are plainly marked by even-aged stands, consisting to the extent of at least 90 per cent of Douglas-fir, regardless of the proportion of Douglas-fir within the original fire-killed stand."

EVERGREEN: What does this mean?
ZYBACH: It means that, as far back as 1740, well before the beginning of white settlement, there is an obvious and measurable history of catastrophic forest fires that have had an impact on wildlife habitat, and it means much of the habitat that was present was in even-age Douglas fir forests the government's scientists apparently believe to be harmful to wildlife.

EVERGREEN: And what does Mr. Munger's writing have to do with the impacts of logging?
ZYBACH: The forests Munger described covered a significant portion of the Douglas-fir region, including most of the habitat that is being designated for spotted owls and marbled murrelets. The nearest modern approximation of prehistoric patterns of fire and reforestation have been large-scale industrial clearcuts that were followed by broadcast burns and successful replantings of Douglas-fir.

These practices produce an almost immediate increase in foliage and protein at the earth's surface, providing important sunlight, habitat and food for many mammals, birds, fish, wildflowers, butterflies and other native flora and fauna. In other words, these forest practices - clearcutting, burning and replanting - result in an immediate increase in plant diversity, which as I understand it, is a major goal of the President's plan.

EVERGREEN: Are you suggesting large-scale clearcuts can be used to approximate forest conditions created by large-scale wildfires and other catastrophic events?
ZYBACH: Clearcutting, broadcast burning and reforestation can be used to closely approximate natural processes by which most old growth stands were first created. Yet there is not a single option in the President's plan that suggests using these tools to rejuvenate diseased or fragmented forests. In fact, these tools are rejected, again revealing a bias against these practices and the people who use them.

EVERGREEN: But wouldn't this sort of logging harm wildlife species that depend on old growth forest habitats?
ZYBACH: This assumes such a cause and effect relationship existed in the past, and there is no evidence in history to support the idea that destruction of old forests, by any means, has caused the extinction of so-called old growth dependent species. Also, it must be remembered that old growth environments present today are much different from old growth environments present when white settlement began. Unfortunately, there is little scientific research that quantifies these differences, and they are almost totally ignored in the President's plan.

EVERGREEN: Are there other important oversights in the President's plan?
ZYBACH: It is apparent the plan's authors don't know much about the history of logging in the northwest, and it shows in what they wrote. There is, for example, a belief that reforestation only became an accepted practice in the northwest 30 or 40 years ago, and a similar belief that concern for overcutting developed in the 1970's with an influx of people fleeing cities.

EVERGREEN: And this isn't true?
ZYBACH: No, it isn't. There was widespread concern for forests extending as far back as 1880 when railroads moving west were consuming about 60 million ties per year. What is interesting about written reports from this time is that the concern was not for old trees, but for millions of acres of young trees that had escaped fire, only to be turned into railroad ties. Repeated references to young trees suggests to me there was a large amount of second growth logging going on before 1900. Survey maps and old photographs support this idea.

EVERGREEN: But were these forests replanted, or were cutover lands simply abandoned?
ZYBACH: Some lands were left in pretty bad shape, but reforestation work began in the northwest long before any of the government's scientists were born. Local planting of Douglas-fir and cottonwood began in the 1850's, and the first commercial planting was done in 1901 by the Willamette Pulp and Paper Company. The U.S. Forest Service started its artificial seeding program in 1908, and the first federal conifer nursery was established in 1910, and tree planting began the same year on the Siuslaw National Forest. By the 1940's, reforestation and forest research work were both firmly established in the northwest. Again, none of this is discussed by the government's scientists, who would have us believe concern for forests is new to the northwest culture. It isn't. Investments in forests and forestry date back to the 1800's.

EVERGREEN: What would you say is the most glaring inadequacy in the President's plan?
ZYBACH: Well, there are many, but I think the greatest inadequacy is revealed in a total lack of understanding of the historic relationship between this region's forests and the people who have lived here for the past 11,000 years.

Several years ago, a man named Hugh Raup, who was a forester and biologist, and taught at Harvard University for many years, wrote something about forest biologists that I believe now explains perfectly why the government's scientists have failed the President and the public. I'd like to read it to you.

EVERGREEN: Please do.
ZYBACK: Mr. Raup wrote, and I quote, "I think the largest single need in American forest biology is the study of man's relation to forest land. Our foresters need to understand much more than most of them do about purely human motives and aspirations with respect to the land. They ought to become genuinely knowledgeable and respectful of people's economic, social and aesthetic institutions."

EVERGREEN: What does Mr. Raup's writing mean to you?
ZYBACH: It means it is impossible to separate land from people. They exist together, as one culture. The government's scientists are trying to separate people from land they and their ancestors have occupied and worked for generations. This is not science. It is cultural engineering, and it is morally and ethically wrong. Moreover, it is environmentally dishonest.

EVERGREEN: What do you think will happen in the region's forests if the President's plan is implemented as proposed?
ZYBACH: I share the concerns of Dr. Oliver and other forest scientists who fear catastrophic wildfire. There is a tremendous amount of dead and dying material in our forests today, a partial result of the long ago made decision to put out wildfires. If these forests are not thinned, you will see wildfires reminiscent of the Tillamook burn, the 1910 fires and the Yellowstone fire. I don't think the public is willing to accept the loss of life and the loss of forests associated with fires this big, and it will not matter to most people that the government's scientists think these fires are "good" because they are "natural."

EVERGREEN: What do you think about the idea that we should return the region's forests to pre-settlement conditions?
ZYBACH: That's a policy question best answered by the general public. But I can tell you this. The pre-settlement conditions the government's scientists have described are not the pre-settlement conditions seen by people who lived here 150 to 200 years ago. Thus, the management plans these scientists want to implement will not produce conditions that existed here when white settlement began. Apart from this failing, I don't find anything wrong with the idea that we should be managing forests in ways that mimic natural processes. Forestry provides the tools we need to do this, but the plan presents a clear bias against these tools, and to mask their bias, they have made up a story about what forests looked like before white men came here. The story isn't true. It is based on a romantic notion that leads people to believe they have few options for protecting forests. That's not true either. We have many, many more options for managing this region's forests than the few that are presented in the President's plan. ■

A section of old growth on Grayback Creek, Siskiyou National Forest, 1993

Historical Research Reveals

The current focus in forest management is aimed at recreating "native" or "natural" landscapes of plants and animals. As such, forest and wildlife managers, both public and private, must deal with an increasing number of laws regarding old-growth, snag retention, endangered species protection, riparian management regulations, logging road construction, and wildfire policies. The intent of these laws is to maintain viable populations of "native" species of plants and animals in their natural (or wild) state. Thus, resource managers are becoming more dependent upon basic understandings of local forest and fire histories for their interpretations of the required and intended practices.

Managing for "native" species presents the question, "Which species really are native?" Many species common to the Northwest such as honeybees, bullfrogs, orchardgrass, opossums, and foxglove have actually been introduced. On the other hand, grizzly bears, timberwolves and California condors were native to western Oregon prior to white settlement but have since become locally extinct. And what about the American Indians? Are they considered a 'native' species? In addition to their biological role at the top of the food chain, they also played a cultural role as firewood gatherers and fieldburners. Now the question becomes, "How many species were dependent upon Indian burning practices for the maintenance of their preferred environments?"

Perhaps the best way to begin answering those questions is to examine historical maps, diaries, photographs, and oral histories of an area. This is exactly what Bob Zybach researched and documented while a graduate student in the College of Forestry at Oregon State University. Zybach has scoured literally tons of evidence which paints a picture of what the Northwest landscape looked like before white settlement. Why is there sudden interest now in his research on pre-white settlement? Because the President's Forest Plan seeks to re-create pre-settlement conditions — those which existed in the early 1800s. *(see story page 7.)*

Zybach's evidence includes photographs, passages from diaries written in the early 1800's by explorers and settlers, notations in journals written by scientists in search of new wonders in a new world, log books from ships which sailed the region's coastline in the 1700's, interviews with Indians whose ancestors inhabited the area, and state and federal forest inventory maps from the 1890's to the early 1900's. All these rich resources reveal a forest history much different than what is promoted in today's media — and the President's Forest Plan.

An attempt to describe natural conditions in the Northwest would require volumes of manuscripts. However, the following descriptions and quotations, selected on the basis of the time, accuracy, and nature of the observation, create a picture of what existed before white settlement, as revealed in Zybach's research.

OREGON AND WASHINGTON COAST RANGES. Available accounts of the indigenous human populations which lived along these coasts begin with a daily journal kept in 1788 by Robert Haswell, an officer aboard Robert Gray's fur trading sloop, *Washington*. Traveling northward from California in nearly constant view of the shoreline, Haswell noted that *"this Country must be thickly inhabited by the many fiers we saw in the night and collums of smoak we would see in the day time,"* and *"a delightful countrey thickly inhabited and cloathed with woods and verdure with maney charming streems of water gushing from the vallies."* By the time the crew had reached the Salmon River, along the central Oregon Coast, Haswell noted that the Indians had *"both Iron and stone knives"* and that *"two or three of our visitors were much pitted with the small pox"*; both certain signs of local European contact and influence.

The settlements and landscapes described by Haswell were also noted by other coastal ship traders in the 1780s and 1790s. The first records of extensive overland explorations along the Oregon coast begin in 1826. During those times, local

Northwest Forests Were Shaped by Fire

Indian settlements were found near the mouths of nearly all the rivers and major creeks which empty into the ocean. Grass prairies and "brakes" (brackenfern prairies), created and maintained by human fires, surrounded most of the Indian settlements and provided habitat for a wide variety of plants and animals.

Writing about conditions in the Tillamook Bay area in 1856, pioneer Warren Vaughn noted:

"At that time there was not a bush or tree to be seen on all those hills, for the Indians kept it burned over every spring, but when the whites came, they stopped the fires for it destroyed the grass and then the young spruces sprang up and grew as we now see them."

Seventy-seven years later, many of those trees fueled the Tillamook Fire of 1933 which burned over 220,000 acres in a single August day, creating a mushroom cloud 40 miles wide that extended eight miles into the atmosphere. When the embers cooled, there was more fuel on the ground than before. Three more fires ravaged that same general area during the next 18 years, often burning hotter than the first fire because all the snags and other fuel had become dry and more flammable.

INTERIOR VALLEYS. The Rogue, Umpqua, and Willamette Valleys in western Oregon and the Puget lowlands of western Washington were occupied by prehistoric peoples who traditionally burned millions of acres of land annually. The result was a nearly contiguous series of great prairies and oak savannahs which extended almost the entire western length of the Cascade Mountains in the US. When the early pioneers headed westward (1835-1848), it was free land in these grassy valleys they sought.

The earliest accounts of prairie burning west of the Cascades dates from 1826, during which year botanist David Douglas traveled with a beaver hunting expedition from Fort Vancouver to the Umpqua Valley. By 1832, most of the Kalapuyan Indians whom Douglas had encountered on his journey had been killed by infectious diseases. Hudson's Bay Company journals from 1825-1834 contain a substantial amount of detailed information which complements and corroborates Douglas' accounts.

By 1842 only a few Kalapuyans lived in western

Continued on page 6

Key Principles in Fire History Findings

Zybach's research findings include a number of basic principles useful for interpreting the evolution and fire history of today's forests:

- **CATASTROPHIC EVENTS.** Forest fuels and human activities in the western US have been shaped by such catastrophic events as volcanic eruptions (Mt. St. Helens), windstorms (Columbus Day Storm), wildfires (the Tillamook Burn), floods, landslides, and climate changes (glaciers, plant migrations, and disappearing lakes and shorelines).

- **HUMAN PRESENCE.** The incremental and cumulative effects of human families — representing the top of the biological food chain — have affected the plants, animals, and fire histories of nearly every acre of Northwest forests during the past 10,000 - 12,000 years.

- **HUMAN ACTIVITIES.** Specific cultural activities, such as firewood gathering and broadcast burning, helped establish local fuel distributions and efficient fire management strategies during prehistoric times.

- **WILDLIFE HABITAT.** Current populations of animal wildlife and patterns of wildlife habitat have been shaped and established by a combination of catastrophic events, climate changes, human presence, and family activities since the time of the last Ice Age.

Continued from page 5

Oregon, but they were still burning and maintaining many of the prairies, wetlands, and oak savannah which provided their families with berries, seeds, nuts, bulbs, weaving fibers, fuel, construction materials and animals. To Indian families an oak tree was a much more valuable species than Douglas-fir because it provided shade in the summer, firewood, protein from the acorns, a hunting perch, and a meeting place. The oak savannahs were maintained by fire up to elevations of about 1500'-2500,' depending on local topographic conditions. Tribes from Chehalis, Vancouver, and Olympia to Sacramento and valleys in between all practiced burning to maintain their environments.

In 1920, George Riddle described his November 1851 impressions of the Umpqua Valley home of these people:

"At that time Cow Creek valley looked like a great wheat field. The Indians, according to their custom, had burned the grass during the summer, and early rains had caused a luxuriant crop of grass on which our immigrant cattle were fat by Christmas time...Fortuneately in our case the land was ready for plow. There was no grubbing to do. In all the low valleys of the Umpqua there was very little undergrowth, the annual fires set by the Indians preventing young growth of timber."

In the same series of articles, Riddle also described another Indian use of fire:

"The massacre of the Indians on Butte Creek occurred on the morning of the 7th of October, 1855. On the 9th and 10th the country between Gold Hill and Galesville on upper Cow creek, a distance of 50 miles, was in a blaze."

According to Zybach, the Indians used fire as a tool to control fire and they did so by removing fuels from the forest floor. "People gathered firewood systematically all along the river banks and mouths, and around the campsites — whatever was closest to the fire locations. In fact, they had canoes out in the streams fishing in logs and moving fuels for trade. In addition, campsites and prairies were seasonally burned to reduce insect populations, to hunt, to clear land and harvest crops. Because these land use patterns resulted in little coarse woody debris, there was often less smoke by today's standards. That's how the Indians controlled fire. They didn't have wildfires, that's a European phenomenon. The Indians were a fire-dependent people who taught by demonstration and word. To say that today we somehow have a better handle on fire is just wrong — and arrogant," Zybach explains.

CASCADE MOUNTAINS.

There are few early contemporary accounts of Indian activities in the Cascade ranges before white settlement. That is mostly because the climate of these mountains is inhospitable to people during much or all of the year. In Oregon, written records of crossing the Cascades begin with beaver hunters in 1825 and 1826. In Washington, the first good records are from botanist William Brackenridge's May 1841 crossing near Mount Rainier. His journals revealed how ladder fuels often developed in wet coniferous forests and how quickly the fires could spread to unoccupied forested areas:

"We now pitched our Tent and there being no water near we melted snow to procure Water for Tea. As the Cooking of the Supper was going on, some one set fire to the moss covered trees, and in one minute all around was one glaring mass of flame. Now was a confusion and bustle — to get all of our luggage to windward, & in the midst of the conflagration Dr. P came near loosing the tails of his Coat by a brand of fire that had accidentally fallen into his pocket."

Modern researchers have used tree rings to help overcome the lack of early written records for the Cascades. Tree rings from old-growth pine on the eastern slope of these mountains show that fire was frequently used by people, burning through some large forests at 5 to 10 year intervals. As a result, early settlers often found a grassy, brush free environment with huge wide spaced trees in which they could ride their horses and graze livestock.

West side Cascade forests were much different, Zybach says. Douglas-fir and hemlock grew instead of pine, and the wetter, steeper slopes were far brushier. With the exception of scattered high elevation meadows and huckleberry patches, there was often little obvious indication of human fire use or occupation.

Let History Do The Talking

Prior to European contact, periodic burning practices were probably the most significant cultural influence on the Pacific northwest environment. Human changes to the environment since then include the introduction of domesticated grazing animals, new human diseases, exotic plants, private property laws (fencing);

metal tools and guns; widespread urban, agricultural and transportation development; organized fire suppression activities; and huge increases in the number of people, their pets, and their visitors.

Current populations of native Northwest plants and animals exist in an environment which has been heavily influenced by the actions of local people, families, and communities through time.

Today's forest plans promote naturally functioning ecosystems defined as those which operate outside of human influence. "Today's ecosystems are a function of Indian disease in the late 1830s, and federal fire suppression in the early 1900s, resulting in coarse woody debris buildup and increase in the incidence of snags. This concept of a naturally functioning ecosystem without human influence is not supported by the historical record," Zybach says.

Today's forest managers must consider that the primary difference in the forests of today, compared with those that existed before the time of European settlement, is the absence of people and their activities from the environment. Perhaps the most telling change is the removal of daily and seasonal fire. If we are truly to return to "pre-settlement conditions," we must first start by reintroducing these elements to ecosystems.

Bob Zybach is currently a forest history and public education specialist with New Albion Multimedia in Corvallis, an ecosystem analysis and curriculum development partnership. He can be reached at 503-757-8387.

Zybach's Research and the President's Forest Plan

*excerpted from an interview with Bob Zybach which appeared in **Evergreen Magazine**, March/April 1994*

E: Mr. Zybach, your criticism of the President's proposal for managing federal forests in the Pacific Northwest has caused quite a stir. Where did you find fault with the plan?
Z: There is nothing in the plan that suggests an understanding of this region's natural and human history. Because many of the plan's goals are founded on erroneous historic assumptions, the plan itself is fundamentally flawed.

E: Are you saying the President's plan is doomed from the start?
Z: It is if its historical basis is not corrected. What is presented as history is both wrong and misleading. To make matters worse, there is a public perception that this plan is based on the best available scientific information. Far from it.

E: What are the errors in history?
Z: There is a poor understanding of the role of fire in Douglas-fir forests. The extent of American Indian involvement in altering forest landscapes is downplayed, and the effects of white settlement, particularly logging, are exaggerated.

E: And these historic errors adversely affect the President's plan?
Z: Yes. How can we expect to re-create pre-settlement forest conditions if we do not fully understand what conditions were present, or why those conditions were present?

E: What are "pre-settlement forest conditions?"
Z: Forest conditions thought to have existed before white settlement began in the early 1800s.

E: And the President's plan seeks to re-create pre-settlement conditions?
Z: Yes.

E: Why?
Z: The plan strongly suggests that the forests which were here before white settlement began were somehow more natural than are present-day forests. To accept this idea, you also have to believe pre-settlement forests were "naturally functioning ecosystems," untouched by human hands. This is their first big error in history. The fact is, people have been altering the character of this region's forests for at least 11,000 years.

E: What would you say is the most glaring inadequacy in the President's plan?
Z: I think the greatest inadequacy is revealed in a total lack of understanding of the historic relationship between this region's forests and the people who have lived here for the past 11,000 years.

Continued on page 16

Fire Improves Forest Health

Continued from page 9

stop that process completely, the landscape gets out of kilter. When that happens, it skews all the forest processes and relationships. This is one reason we're seeing all these forest health problems — because the landscape has moved to one extreme and is unstable. We think our efforts in the Santiam Vegetation Forest Health Project will result in a better long-term balance.

EDITOR'S NOTE: *Two appeals on the Santiam Vegetation Forest Health Project EIS were filed during the appeal period which just ended.*

Filed by the Oregon Natural Resources Council and the Sierra Club, the appeals are for the entire project, not any specific element. At this time, the Forest Service is preparing a written response to the appellants which will include a decision as to whether or not the appeals are denied.

If they are denied, the appellants may still go to federal court and ask for an injunction.

The Research and the President's Forest Plan

Continued from page 7

Several years ago, a man named Hugh Raup, who was a forester and biologist, and taught at Harvard University for many years, wrote something about forest biologists that I believe now explains perfectly why the government's scientists have failed the President and the public. Let me read it to you.

"I think the largest single need in American forest biology is the study of man's relationship to forest land. Our foresters need to understand much more than most of them do about purely human motives and aspirations with respect to the land. They ought to become genuinely knowledgeable and respectful of people's economic, social and aesthetic institutions."

E: What does that mean to you?

Z: It means it is impossible to separate land from people. They exist together, as one culture. The government's scientists are trying to separate people from land they and their ancestors have occupied and worked for generations. This is not science. It is cultural engineering, and it is morally and ethically wrong. Moreover, it is environmentally dishonest.

E: What do you think about the idea that we should return the region's forests to pre-settlement conditions?

Z: That's a policy question best answered by the general public. But I can tell you this. The pre-settlement conditions the government's scientists have described are not the pre-settlement conditions seen by people who lived here 150-200 years ago. Thus, the management plans these scientists want to implement will not produce conditions which existed here when white settlement began. Apart from this failing, I don't find anything wrong with the idea that we should be managing forests in ways which mimic natural processes. Forestry provides the tools we need to do this, but the plan presents a clear bias against these tools, and to mask their bias, they have made up a story about what forests looked like before white men came here. The story isn't true. It is based on a romantic notion which leads people to believe they have few options for protecting forests. That's not true either. We have many, many more options for managing this region's forests than the few that are presented in the President's plan.

Forests

TODAY & FOREVER

P.O. Box 40184 • Eugene, OR 97404-0025

NONPROFIT
U.S. Postage
PAID
Eugene, OR
Permit # 627

Advances in Fire Practice

Wildland Fire Lessons Learned Center

U.S. Wildfire Cost-Plus-Loss Economics Project: The "One-Pager" Checklist

By Bob Zybach, Michael Dubrasich, Gregory Brenner, John Marker

FALL 2009

Topics: Economics, Fire Management, Research Resources

What are the actual costs of a wildfire?

Official Forest Service tallies usually include suppression expenses only. Media reports sometimes include estimates of damage to homes and infrastructure. But the economic impacts of wildfires are far-reaching and new (and old) research shows the need for improved cost estimates of wildfire.

Large wildfires consume more than just suppression expenses ("costs") – they also do measurable short- and long-term damages ("loss") to public and private equity and resources. Traditional fire appraisal uses the term "cost-plus-loss" to account for all the economic impacts of wildfire. This econometric analysis method is sometimes expressed as LCD (least cost plus damage) or C+NVC (costs plus net value change). The goal (economic utility) of fire suppression is to minimize cost-plus-loss.

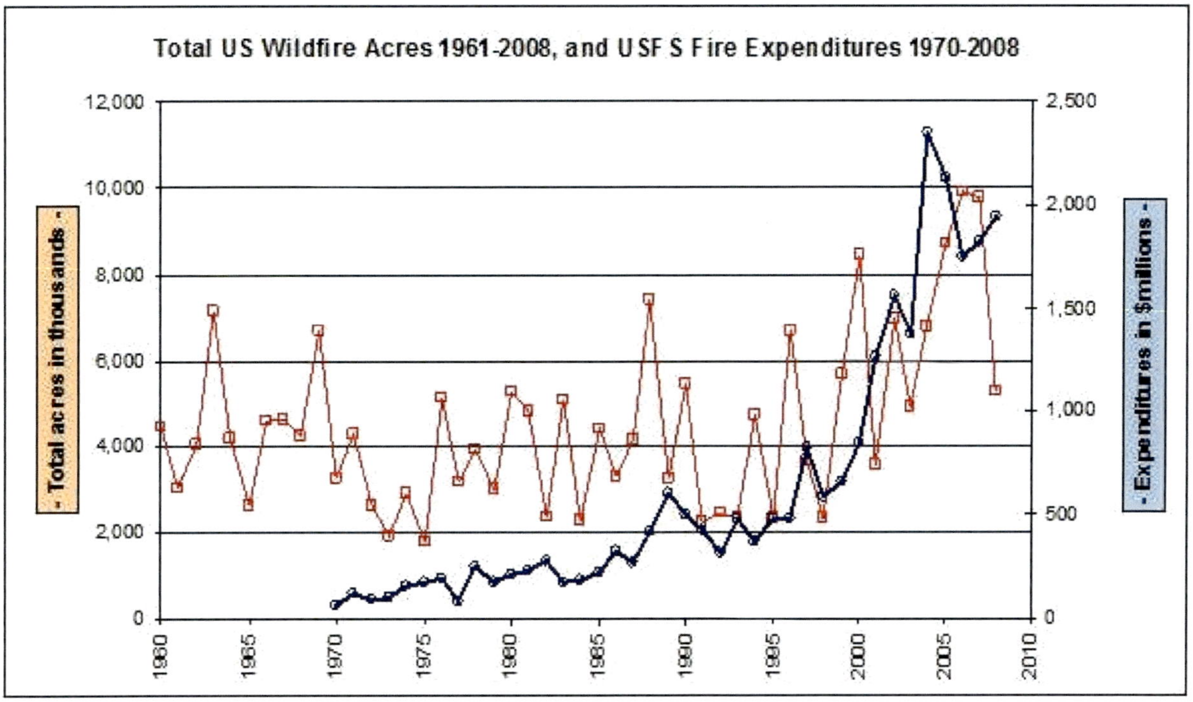

Recently analysts, government officials, and the media have drawn increasing attention to the escalating frequency, severity, and costs over and above fire suppression associated with large-scale forest wildfires [1] – including losses of human lives, homes, pets, crops, livestock and environmental damage.

- The Western Forestry Leadership Coalition recently released a report entitled "The True Cost of Wildfire in the Western U.S." (Dale et al 2009). The authors examined six major US wildfires, and compared suppression costs and tactics with "total costs." Two examples of this process were the 2000 Cerro Grande fire in New Mexico (shown to have suppression costs that reflected only 3% of total damage estimates), and the 2003 Old, Grand Prix, and Padua fire complex in California, in which suppression costs were only 7% of total costs to date – with total losses expected to increase dramatically in years to come (Dunn et al, 2005).

- The 2003 fires in San Diego and Southern California were a disaster by any measure – 24 fatalities, over 3,700 homes destroyed. At the time, the costs of the suppression efforts were staggering, $43 million. However, Matt Rahn, a researcher from San Diego State University, recently presented findings that put this figure at less than 2% of the total long-term cost of the fire (Rahn, 2009).

- The Hayman Fire (2002) burned 138,000 acres and cost $42,279,000 ($307/acre) to suppress. But Professor Dennis Lynch of Colorado State University estimated that an additional $187,500,000 ($1,358/acre) in losses had accrued within a year. Suppression costs were only 18% of the total, and Dr. Lynch stated, "I recognized the need to follow costs into subsequent years to more completely identify a fire's true impact" (Lynch, 2004).

To date, our own findings paint a far different picture than that commonly reported by the media or understood by the public. We have found that total short-term and long-term cost-plus-loss attributed to wildfires typically attains amounts that are ten to 50 times (or more) reported suppression expenses.

The Pacific Crest Trail, Mount Jefferson Wilderness Area, 2004. In addition to being unsightly and dangerous due to the threat of falling limbs, trees and reburning, much of this trail segment has been closed or difficult to traverse since the 2003 B&B Fire Complex. (Photo: B. Zybach).

Losses in excess of suppression costs include economic damages to timber and forage values, wildlife habitat and populations (including endangered species and their critically protected habitat), air and water quality, recreational opportunities, public health, private businesses, and other resources and amenities important to all citizens. The National Association of Forest Service Retirees issued "Forest Health and Fire: An Overview and Evaluation" (Pfilf et al., 2002) that documented and analyzed the recent historic increases in US wildfire occurrences and severity. The report called for a detailed accounting of "total losses associated with fire and other forest health situations," specifically mentioning homes, evacuations, insurance claims, natural resources, recreation, water, forest health, timber, habitat, wildlife, management costs, subsequent increases of insects and disease.

Appraisal of resources damaged by wildfire is not always straightforward. Human lives and adverse health effects are usually not considered in terms of dollar losses at all, and tallies of wildlife fatalities are rarely done. Habitat, scenery, ecosystem services, and other non-commodities are difficult to value, although there has been considerable study and published efforts in that regard. Rarely has there been any attempt to quantify the long-term consequences of a damaged renewable resource base to provide for the needs of an ever increasing present and future human society (e.g. Bowman et al. 2009).

Consideration of an inclusive and comprehensive cost-plus-loss evaluation could be a helpful exercise when evaluating suppression/readiness need and effectiveness appropriation. We offer an operational methodology to better characterize the broader true costs to society of large-scale wildfires: typically, those fires involving loss of human life, thousands of acres in size, and/or millions of dollars of damage. This article focuses on practical uses of a "first step" tool we have developed – a one-page checklist – that can be employed by local citizens, media, fire managers, and elected officials to begin the process of better accounting for short-term and long-term effects of wildfires on US lives and economies. We use comprehensive "cost-plus-loss" accounting methods and consider direct, indirect, and post-fire costs and losses. We have developed eleven separate ledger categories of costs and losses that are reflected in the general ledger, and these same categories are listed in the "one-pager" checklist (see Table 1), as further described in the following. Within each ledger category, costs and losses are characterized as direct, indirect, and post-fire.

Direct costs. These are the amounts spent directly controlling a wildfire (suppression) and wildfire-related expenses, such as evacuations, commerce disruption, equipment damage, burnt homes, cars, and personal property, school and playground closures, additional air quality monitoring, public health alerts, or other costs directly related (and generally concurrent) to the fire.

Indirect costs. These have been typically over-looked in accounting for wildfire damages in earlier reports. Indirect costs include amortized wildfire preparedness expenditures such as crew training costs, equipment and supply outlays and depletion planning, and fire insurance premiums. Damage to capital investments such as recreational structures, devalued experiences, investments in forest management (reforestation, thinning) agriculture (crop establishment and treatments), past property taxes, reduced air and water quality, and changed landscape aesthetics are other indirect costs. This category considers indirect costs that are concurrent to a wildfire; i.e., accrued as the fire occurs. [2]

Post-fire costs. These are the long-term damages (losses), direct and indirect, to society and the environment. These include capital value losses to timber, agriculture, homes, and other public and private equity. Some post-fire losses can be difficult to quantify, and may only become apparent over time. Long-term human health effects, increased costs of medical care, reduced property values due to wildfire smoke damage, rehabilitation costs for publicly and privately damaged facilities, negative impacts on affected livelihoods, and sediment management costs in reservoirs subjected to increased soil erosion are examples of post-fire costs that may be attributed to specific wildfire events. Yet to be fully documented are wildfire smoke emission effects on possible climate change.

These are the eleven categories, with basic definitions, that are used in the checklist "one-pager," and in the comprehensive ledger:

1. Suppression costs. These costs are the ones most commonly reported by media (to the exclusion of other costs and losses) and are often under-reported at that. Typical costs include wages, transportation, equipment, services, supplies, etc. Special costs, such as equipment depreciation, communications interruptions, and emergency evacuations, need to be accounted for, as well. Indirect suppression costs include emergency preparedness measures, supply purchases, crew training, and equipment maintenance. Post-fire costs and losses include equipment repair, supplies replacement, formal reviews, and possible medical treatments and hospitalization of personnel.

2. Property Damage and destruction to federal, state, county, private, and municipal structures and facilities are major losses attributed to wildfires, particularly when transportation networks and communications systems are considered. Damage to timber and agricultural crops are other direct property losses. Some structural losses to private property may be insured; these can include business properties and homes, vehicles, and livestock. Capital goods and equipment damage and depreciation, evacuation expenses, and other losses are directly related to fire and smoke damage. Indirect losses include pre-fire insurance premium payments, building and landscape maintenance expenses, firefighting equipment purchases, and fire-related business closures. Post-fire losses include salvage, clean-up, rehabilitation, and repair expenses, equipment and capital goods replacement, drinking water pollution, smoke damage, deflated real estate values, lost sales tax revenues, and fire insurance premium increases.

3. Public health. These are some of the most overlooked and potentially costly areas associated with wildfire (and resultant smoke) damage. In addition to fatalities, wildfire smoke inhalation is known to cause and exacerbate a wide range of human health problems, including asthma, emphysema, and heart disease. Medical equipment, health-related evacuations, ambulance charges, and hospitalization are some of the direct health losses related to wildfire. Indirect losses include health insurance premiums, pre-fire medical equipment purchases, and medical personnel training. Post-fire losses include long-term health effects and increased health care expenses, insurance premium adjustments, health-related work absenteeism, survivor benefits, and even funeral and burial costs.

4. Vegetation. Standing timber losses have often been considered in wildfire damage estimates, but loss of future harvests from destruction of growing stock has been less frequently accounted for. On public lands losses include destruction of forage on grazing lands, secondary forest products destruction and/or degradation, and loss or degradation of wildlife habitat (including endangered species populations and protected habitats). On private land vegetation losses include timber and agricultural crops burned or impacted by wildfire smoke, such as wine grapes. Indirect losses include mortality of growing stock, the pre-fire investments used to establish or maintain such stock (such as nursery or planting costs and fencing), and irrigation systems. Post-fire losses include seeding, planting, and other revegetation costs, landscape rehabilitation, lost timber growth, and related product sales, business, job, and tax losses.

5. Wildlife. In addition to mortality of forest, range, and aquatic wildlife populations, direct losses include damage and destruction to a wide variety of common or protected habitats and to such amenities as viewing areas and feeding stations. Indirect losses include damage and destruction to pre-fire habitat improvement projects, population enhancement costs, and investments in wildlife research. Post-fire costs include reduced population productivity, foregone game management income, habitat restoration expenses, and related business, job, and tax losses. The loss of listed endangered animals and their habitat is included here.

6. Water. Direct losses include water usage for suppression action, local water system shutdowns, and reductions in drinking water, hydropower, and irrigation supplies and sales. Indirect losses are related to pre-fire planning, system investments, and wildfire-related pollution control devices. Post-fire losses include degradation of domestic water, irrigation, and hydropower supplies, system repairs, administrative costs, sediment and pollution controls and mitigation, and long-term changes in water yield and watershed ability to collect and store water.

Tillamook Fire smoke plume, August 24, 1933. In a single day the 1933 Tillamook Fire increased more than 200,000-acres in size, creating a mushroom cloud 40 miles wide and 8 miles high and producing hurricane-scale winds that furthered the spread of the fire. This cloud was formed largely by water vapor, ash, soil, carbon monoxide, and carbon dioxide in nearly immeasurable amounts and proportions (Photo: The Oregonian, courtesy of Oregon Department of Forestry).

7. Air and atmospheric effects. Direct losses are related to air pollution, including particulate, noxious gases, and CO_2 emissions, and visibility impacts to road and air transportation, especially if delays and/or accidents result. Indirect losses are related to public health effects, property damage, and compromised recreational opportunities.

Post-fire losses include additional air pollution controls, carbon mitigation costs, added administrative overhead, and future reductions in business, job, and tax revenues.

8. Soil-related effects. Soil erosion can occur during a wildfire due to fire-induced wind, or from suppression actions. Soils can be baked, as well. Indirect losses include investments in fertilization, scientific research, and planning. Post-fire losses include decreased soil productivity, increased soil erosion, and post-fire soil rehabilitation, erosion and sediment mitigation, and project administration.

9. Recreation and aesthetics. Campground closures, evacuations, recreation-related business shutdowns, and structural assets damages and destruction are direct recreation losses attributable to wildfires. Indirect losses include pre-fire recreation-related investments by agencies, businesses, and individuals. Post-fire losses include recreational activity decline, degradation of scenic values, compromised hunting, fishing, hiking, camping, and wildlife viewing experiences, recreation-related structural repair or rehabilitation, and reduced business income, jobs, and tax revenues.

10. Energy. Direct losses include transmission line shutdowns and resultant loss of metered power sales, destruction and damage to energy production and transmission systems, and loss of biomass energy supplies. Indirect losses include pre-fire investments in energy production facilities and transmission systems and power planning costs. Post-fire losses include energy sales reductions, equipment repair, added sediment control, and future business, job, and tax revenue losses.

11. Heritage (cultural and historical resources). These losses include damage and destruction of historical resources and pre-Contact archaeological sites, loss or damage to historic cultural trail systems, ceremonial sites, and sacred sites, and heritage-related business shutdowns. Indirect losses include pre-fire public and private investments in heritage resources, including formal evaluations, research, and structural improvements. Post-fire losses include heritage site rehabilitation and repair costs, devaluation of cultural and spiritual assets, the loss of traditional uses and heritage, lost research opportunities to gather limited and fragile information, and heritage-related business, job, and tax revenue declines.

Conclusions

US wildfire events have become increasingly large, destructive, and costly during the past 20 years, and particularly since the turn of the century. During this time wildfire suppression costs have also increased dramatically. Suppression costs, however, represent only a small fraction of over-all wildfire cost-plus-loss. Other concurrent direct and indirect losses together with long-term post-fire losses can total 10 to 50 times (or more) the suppression costs. A more comprehensive economic and risk analysis and awareness on the part of decision-makers and the public of wildfire cost-plus-losses is needed, as are land and property management reforms, to help reverse these trends.

Our one-page checklist is intended to make initial estimates of total fire costs, and to ultimately be used in conjunction with a more comprehensive ledger for better tracking costs and losses over time. We believe that the use of these tools will better inform land

and resource managers in the management of fuels and wildfires by identifying true costs of decisions and allowing better judgment in the establishment of resource use priorities.

How to Use the Cost-Plus-Loss Checklist

The checklist is meant to help expand the discussion regarding the overall, long-term impacts of wildfire. Use the category descriptions to start the search for information regarding wider wildfire impacts on areas of concern in your community. Fully accounting for impacts under each of these categories is difficult, as data and information are often difficult to obtain, but some is available with a little digging.

1. Start by checking off all those items that don't apply.
2. Next, insert readily available numbers, particularly those quoting "experts" in the local media. Human fatalities, acres, and other data can also be acquired in this manner.
3. Insert "placeholder" estimates for precise numbers becoming known — such as final suppression costs, insurance payments, etc. This sub-total would be near the lowest end range of values for damage from a particular event.
4. Insert estimates for unknown or long-term costs, such as wildlife habitat loss, public health effects, air pollution, etc. Contact local experts for estimates, or even rough estimates in a range of values.
5. Continue to update list as better information becomes available — whether via existing sources, research, or a combination of the two utilizing the general ledger.

References

Bonnicksen, Thomas M. 2008. The Forest Carbon And Emissions Model. The Forest Foundation, Auburn, CA: Reports 1 & 2, Overview and Technical Information - 28 p., Greenhouse Gas Emissions From Four California Wildfires: Opportunities to Prevent and Reverse Environmental and Climate Impacts - 19 p.

Bowman, David M., Jennifer K. Balch, Paulo Artaxo, William J. Bond, Jean M. Carlson, Mark A. Cochrane, Carla M. D'Antonio, Ruth S. DeFries, John C. Doyle, Sandy P. Harrison, Fay H. Johnston, Jon E. Keeley, Meg A. Krawchuk, Christian A. Kull, Brad Marston, Max A. Moritz, I. Colin Prentice, Christopher I. Roos, Andrew C. Scott, Thomas W. Swetnam, Guido R. van der Werf, and Stephen J. Pyne 2009. "Fire in the Earth System," Science, Vol. 324, No. 5926: 481 – 484.

Dale, Lisa 2009. The True Cost of Wildfire in the Western U.S. Western Forestry Leadership Coalition, Lakewood, Colorado: 16 pp.

Dubrasich, Michael, Bob Zybach, Greg Brenner, and John Marker 2009. U.S. Wildfire Cost-Plus-Loss Economics Project: Long-form Accounting Ledger. In Review.

Lynch, Dennis L. 2004. "What do Forest Fires Really Cost?" Journal of Forestry, September 2004.

Pfilf, Richard J., John F. Marker, and Robert D. Averill 2002. Forest Health and Fire: An Overview and Evaluation. National Association of Forest Service Retirees, Chantilly, Virginia: 36 pp.

---

[1]It is beyond the scope of this article to examine or discuss the possible causes of recent escalations in wildfire occurrences, suppression costs, or cumulative damages – only to acknowledge these facts, which are well documented in numerous publications and other media.

[2] Although we have standardized costs to dollar figures, values place on human lives, aesthetics, or pollution, as examples, are often highly subjective and difficult or impossible to quantify. Their inclusion in the checklist is simply to acknowledge their existence, and to allow for users to better consider such values in context with other wildfire impacts. A more specific example is provided by current concern with greenhouse gas emissions related to wildfire smoke. Thomas M. Bonnicksen has developed a mathematical model used for estimating the amount of greenhouse gases (CO_2, CH_4, N_2O) emitted by forest fires (Bonnicksen 2009). Using his model, he estimated that the 66,000-acre Moonlight Fire (2007) on the Plumas National Forest generated nearly 5 million tons of greenhouse gases. How can this amount be expressed in dollar terms? Will cap-and-trade legislation currently before the US Congress affect such an evaluation? The same model was used to calculate that 40 million tons of greenhouse gases were released or will be released as the dead wood decays that resulted from the 800,000 acre Idaho Batholith fires of 2007, and that the 1.34 million acres that burned in wildfires in California in 2008 generated CO_2 emissions equivalent to approximately 13 million cars driven all year. The checklist is not intended to answer these questions, but to account for them and put them in context.

Table 1. Wildfire 'Cost-Plus-Loss' Ledger Checklist Form

Fire Name ____________________ County __________ State _____ Country __________
Ignition Date _____________ Containment Date _____________ Total Acres __________
Cause: Human___, Lightning__, Operation__, Prescription__, Maintenance__, Other___
Major Landowner(s) ___________________ Human Fatalities ____ Homes Lost _______

Cost-Plus-Loss Category	A. Direct	B. Indirect	C. Post Fire	Totals
1. Suppression Costs a. Public				
b. Tribal/Private				
2. Property Damage a. Public				
b. Tribal/Private				
3. Health Effects a. Public				
b. Tribal/Private				
4. Vegetation a. Public				
b. Tribal/Private				
5. Wildlife a. Public				
b. Tribal/Private				
6. Water a. Public				
b. Tribal/Private				
7. Air and Atmospheric a. Public				
b. Tribal/Private				
8. Soil-Related a. Public				
b. Tribal/Private				
9. Recreation a. Public				
b. Tribal/Private				
10. Energy a. Public				
b. Tribal/Private				
11. Heritage a. Public				
b. Tribal/Private				
Totals				

Name ____________________________ Title ____________________ Affiliation _________________

Date ______________________________

Predicting Oregon's 2012 Major Wildfire Events

By Bob Zybach, PhD

Because my PhD is in the study of historic catastrophic wildfires in western Oregon – and because I am part of a network of other scientists, professionals, and resource managers with similar interests – every year about this time I seem to get asked the same basic questions: "Bob, what do you think the chances are of Oregon having a really huge wildfire this year?" And, "When and where can we best expect it/them to take place if and when they do occur?"

The fact is I have no idea what the answer is to those questions. Nobody wants to hear facts, so my given answer always is: "August and September will be the riskiest months west of the Cascade crest (see Fig. 1), with likelihood of uncontrollable wildfire to increase in proportion to length in time of east wind events, and severity in proportion to condition, continuity, and volume of available fuels. Eastside is basically the same for forested areas and shrublands, with the riskiest months beginning in July and ending in October."

Fig. 1. Smoke plume, Rainbow Creek Fire, September 22, 2009. This fire became more than 3,000-acres within 24 hours, and grew to more than 6,000-acres total. The Boze Fire, in the foreground, became more than 10,000-acres total (Photo by J. Baldwin, US Forest Service, obtained online: www.inciweb.org/incident/1893/).

Most of the time my predictions are fairly accurate, depending on how one defines "risky," and whether any prolonged east wind events take place during those months for that year. Other than vague adjectives and qualifiers, however, the main reason for my accuracy is that it is based on a generalized description of almost all of the major wildfire events that have taken place since Americans first began emigrating to the Oregon Country in the early 1840s (see Fig. 2). There is a good documentary record of the "Great Fires" from that time until now, and they are basically late summer-early fall events, defined in large part by east winds.

Realistically, it is difficult to tell when a wildfire is going to take place, and from there it becomes a matter of hours and minutes as to how it is affected by – and how it directly affects – local weather conditions. Wildfires can create gale-force winds, lightning, tornadoes, cumulus clouds, and even rain under the right conditions, and almost all of these occurrences are unpredictable even minutes before they begin taking place.

Fig. 2. Tillamook smoke plume over the Willamette Valley, August 24, 1933. On this single day in 1933, the Tillamook Fire increased more than 200,000-acres in size, creating a mushroom cloud 40 miles wide and 8 miles high and producing hurricane-scale winds that furthered the spread of the fire. (Photo from Portland, Oregon Oregonian, used by permission of the Oregon Department of Forestry).

The "volume of fuels" is another matter; we know where they are and what they are – it's just whether or not they will burn in a wildfire, and if so: to what degree and when? Pitchy conifer trees, snags (standing dead trees), and downed wood are among the most volatile wildfire fuels that exist. Oregon contains among the largest contiguous amounts of these types of fuel in the world. The potential severity of this condition is increased as these fuels have increased in both volume and in total acreage during the past 30 years. This phenomenon is particularly apparent in the ancient prairies, savannahs, berry fields, and pine woodlands that have become invaded with Douglas-fir, lodgepole, true fir, and juniper during historical time -- and also true for vast acreages of the State that have become covered with millions of snags in recent years; victims of bugs, wildfire, competition, and passive forest management (see Fig. 3).

The likelihood and severity ("plant and animal mortality") of a wildfire, then, is dependent on what starts burning and where; what the weather is like in the immediate vicinity of the fire; and how the fuel is arranged across the landscape. Fires typically – but not always -- move upslope faster and hotter than down slope or across a flat, and move easier through fuels that are continuous, rather than broken by openings or abrupt changes in vegetation types or fuel moisture.

Fig. 3. Pacific Crest Trail, near the Santiam Pass, May 14, 2004. The August-September, 2003 B&B Complex was the largest single wildfire in Oregon Cascades history, totaling more than 95,000-acres in size. As snows began to melt the following spring, it became apparent what damage had been done to popular hiking trails, campgrounds, scenic viewpoints, and other recreational opportunities (Photo by B. Zybach).

The final ingredient is fire. Almost every wildfire is started by either people or lightning, so if we can predict where they will be between July and October, we can better predict where wildfires (and particularly, big wildfires) are most apt to take place. Lightning-caused fire is fairly predictable (see Map 1). It is people – the only animal capable of making and using fire – that are unpredictable. Almost every one of us uses fire in some form or another nearly every day; and most of us make poor or illogical decisions from time to time as well. The vast majority of wildfires are made by people making poor or illogical decisions with fire at unpredictable times in unpredictable locations (see Fig. 4).

That being said, the intent of this article is to provide readers with the best idea as to whether Oregon will experience another catastrophic-scale wildfire (typically, 100,000-acres or more in size), or not, in 2012; and if so, when and where are such events most likely to take place? Of more common concern: when and where are any wildfires likely to occur this year?

To answer those questions, and in addition to my own generalized prediction, I've assembled three predictive maps to compare with what really does happen this fire season. Three months from now, in the next issue of Oregon Fish & Wildlife Journal, we will revisit these predictions and see which were most accurate – just remember: each is predicting wildfire within different parameters, and each comes with its own qualifying jargon. With proper syntax, it is possible that all four projections will be accurate. Or not.

Fig. 4. Burned home, Deer Creek Fire, August 28, 2005. Although this fire near Selma was a relatively small 1,600- acres, several families lost their homes because of it, and it was the largest wildfire in Oregon that year (Photo by B. Zybach).

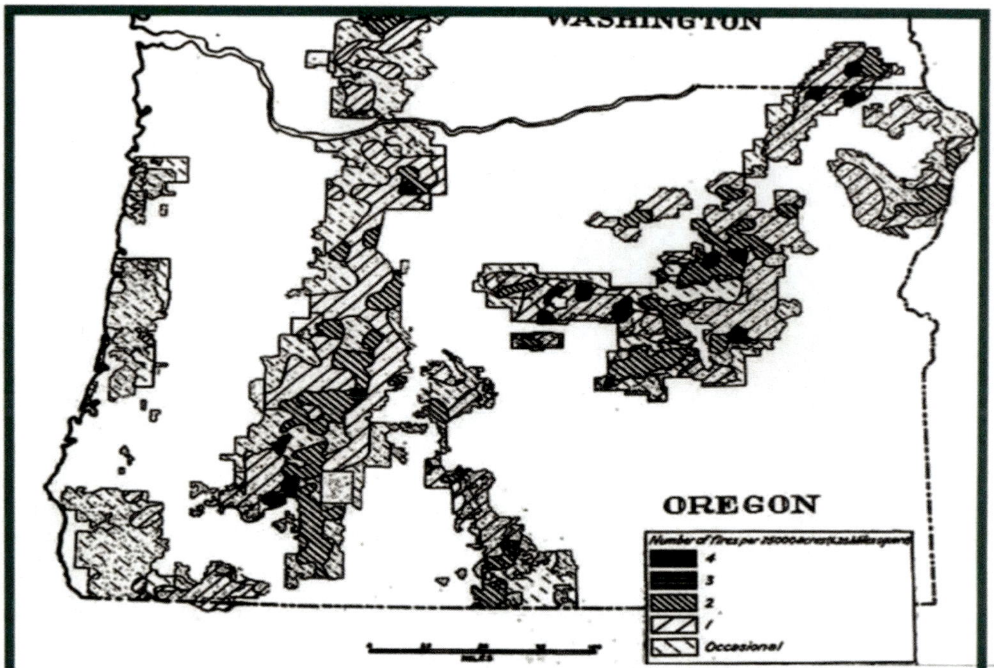

FIGURE 14. – Zones of average yearly lightning fire distribution on the national forests of Oregon and Washington obtained by plotting the locations of the 5300 lightning fires reported from 1925 to 1931

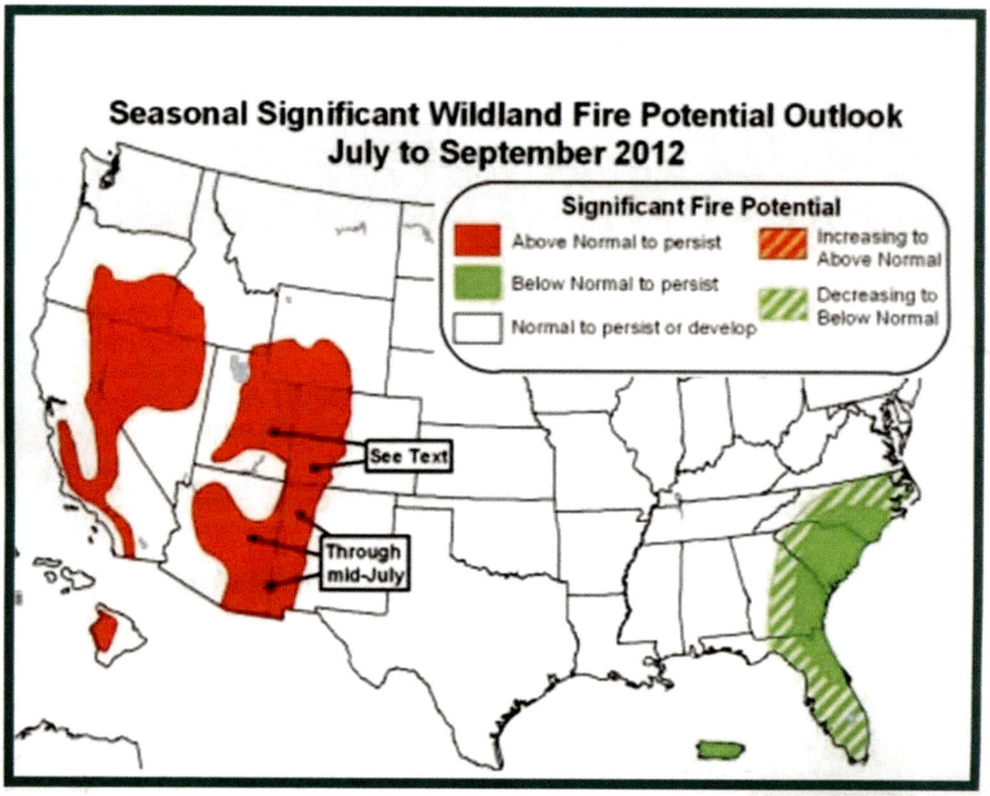

Map 1 was put together in 1934 by an expert on wildfires, William G. Morris, so as to better prepare wildfire managers to deal with future events. At that time – near the beginning of the Dust Bowl 1930s -- there was no concern about climate change; assumptions were that weather patterns would remain about the same through time, while acknowledging floods, droughts, freezes, windstorms, etc., "of biblical proportions" from time to time. Morris' map shows "Occasional" lightning fires along the Oregon Coast, but increasing numbers of such fires along the entire length of the Cascades, northeasterly through the Ochocos and Blues to Hells Canyon, and southeasterly along the northwestern rim of the Great Basin to California.

Map 2 was issued on June 1, 2012 by Predictive Services for the National Interagency Fire Council in Boise, Idaho. This is their job. They have divided the United States into five different classifications for the next three months, with Oregon given two classifications: Normal (white) and Above Normal (red). Of particular interest is that the entire red area is contained in the one part of Oregon that doesn't drain into the Pacific Ocean – the Great Basin. On the surface, a geographical area that is best known for salt lakes and vast deserts with hardly any vegetation would seemingly be an unlikely choice as the part of the State most likely to experience a wildfire; but that portion of the Basin in southeast Oregon also features the Steens Mountains and numerous forested areas, shrublands, and grasslands that have grown rank with invasive native plants and exotic weeds in recent decades. Too, the July to September timeline fits in nicely with my own July to October prediction, although I safely included everything east of the Cascades, not just the Great Basin lands.

Map 3 was constructed June 8-9, 2012 with the direct help and advice of the people who developed and maintain this particular predictive model for the entire US. It represents the most current results from a vegetation/wildfire predictive model first assembled by James Lenihan, US Forest Service, Pacific Northwest Research Station (MAPSS Team, Ron Neilson team leader - now retired), in the late 1990s; with five climate projections provided by the International Research Institute for Climate and Society formatted by Ray Drapek (MAPSS team); to the transformation of his output data to GIS layers (http://bit.ly/JU39Zy) for storage, display, and delivery (this map, as an example) by Dominique Bachelet and Conservation Biology Institute staff; through Data Basin (http://databasin.org/).

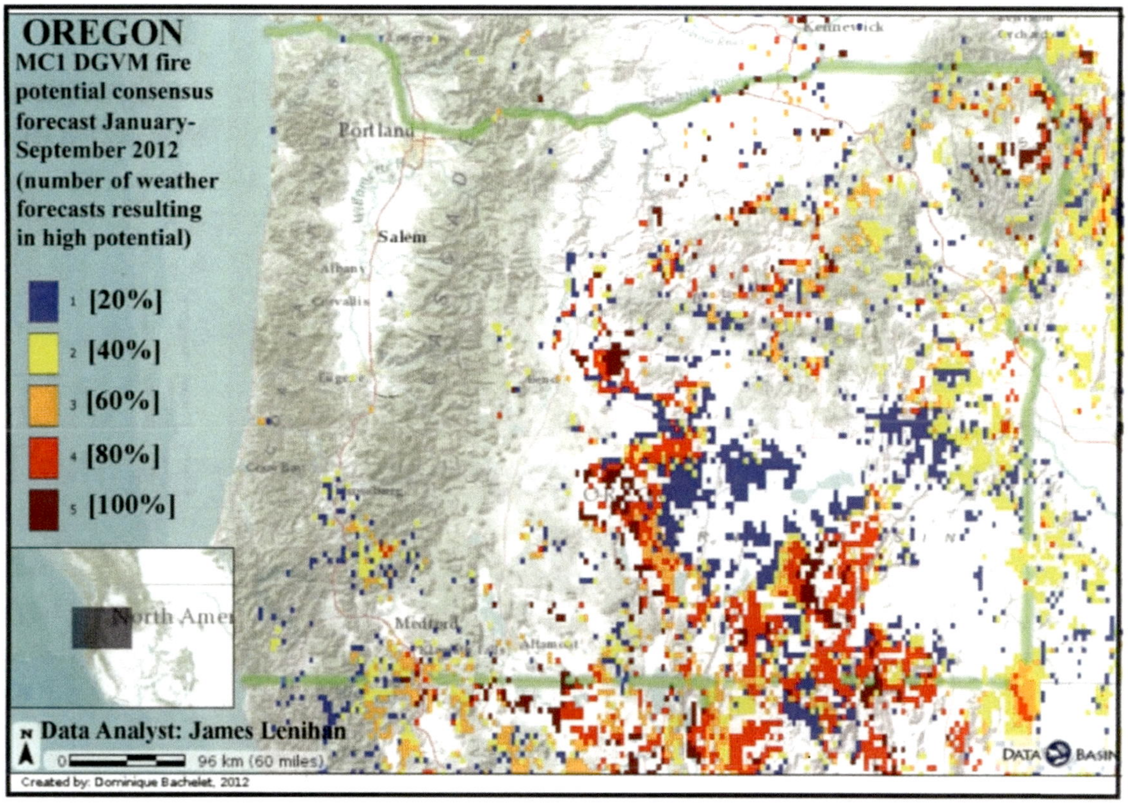

So, that's all the fine print, and also useful links for those readers with a technical or professional interest in these topics. The take home message is Map 3 colors represent the "percentage agreement between the five projected fire risk maps driven by five different weather forecasts." That is, Blue shows that "under one" of the five future climates (>20%) the fire forecast model simulates a potential future wildfire problem area; if the pixel is orange, then under three future climates (>60%) fire forecast results are in agreement, and if it is maroon, all five climates (100%) cause the fire forecast model to show that particular area as most likely to experience wildfire this year.

What does Map 3 tell us? First, there is little agreement among the five climate drivers on wildfire danger this summer west of the Cascades – mostly along the I-5 corridor south and east of Roseburg, to the California border, with greatest likelihood along the Klamath and Applegate rivers. On the east side, a vortex of high risks seems to form to the northeast of Bend, near Prineville: with one arm extending northeasterly through the Ochocos and Blue Mountains, to a heavy concentration of maroon pixels in the Wallowa-Whitman National Forest on the west side of Hells Canyon; and with another arm moving southeasterly, along the northwestern boundary of the Great Basin in the Fremont National Forest. Surprisingly little activity predicted for the Deschutes, Mt. Hood, Willamette, and Rogue-Siskiyou National Forests, among others.

Conclusions

In many ways, Map 3 agrees with Map 2 in that both predict the greatest likelihood of wildfires in Oregon this fire season to be in the Great Basin sometime between July and September. Map 3 is more detailed, however, and excludes the Alvord Desert and other portions of southeast Oregon that don't have enough fuel to create a wildfire in the first place. Another surprising consistency between the maps is that neither indicates much risk (other than "normal") for western Oregon and for the eastern slope of the Cascades. Maybe the definition for "normal," though, is becoming modified by such historic events as the Biscuit Fire and B&B Complex.

Wildfires are started by people or lightning. Map 1 shows the history of more than 5,000 reported lightning strike wildfires throughout Oregon from 1927-1932. If the historical record is a reasonable predictor of future events, then this map should also be considered until these patterns can be shown to have changed. Two things of note: there is considerably more risk shown along the length of the Cascades than indicated by either of the other maps; and there is nothing shown for the Great Basin. In the first regard, perhaps Map 1 is simply depicting the "normal" wildfire risk for those areas; and so far as southeastern Oregon is concerned, much of it was still being heavily grazed by sheep and regularly burned by sheepherders when this map was made –and probably nobody bothered to "report" a wildfire in that area at that time anyway. No one cared; now we have satellites.

So there we have it: four different predictions for Oregon's coming fire season, three documented with maps and one generalized with qualifiers (but an exacting timeframe). In the next issue of Oregon Fish & Wildlife Journal we'll see whom, if anyone, best predicted this year's wildfire events.

Predictable And Preventable!

Oregon's 2012 Wildfires

By Dr. Bob Zybach

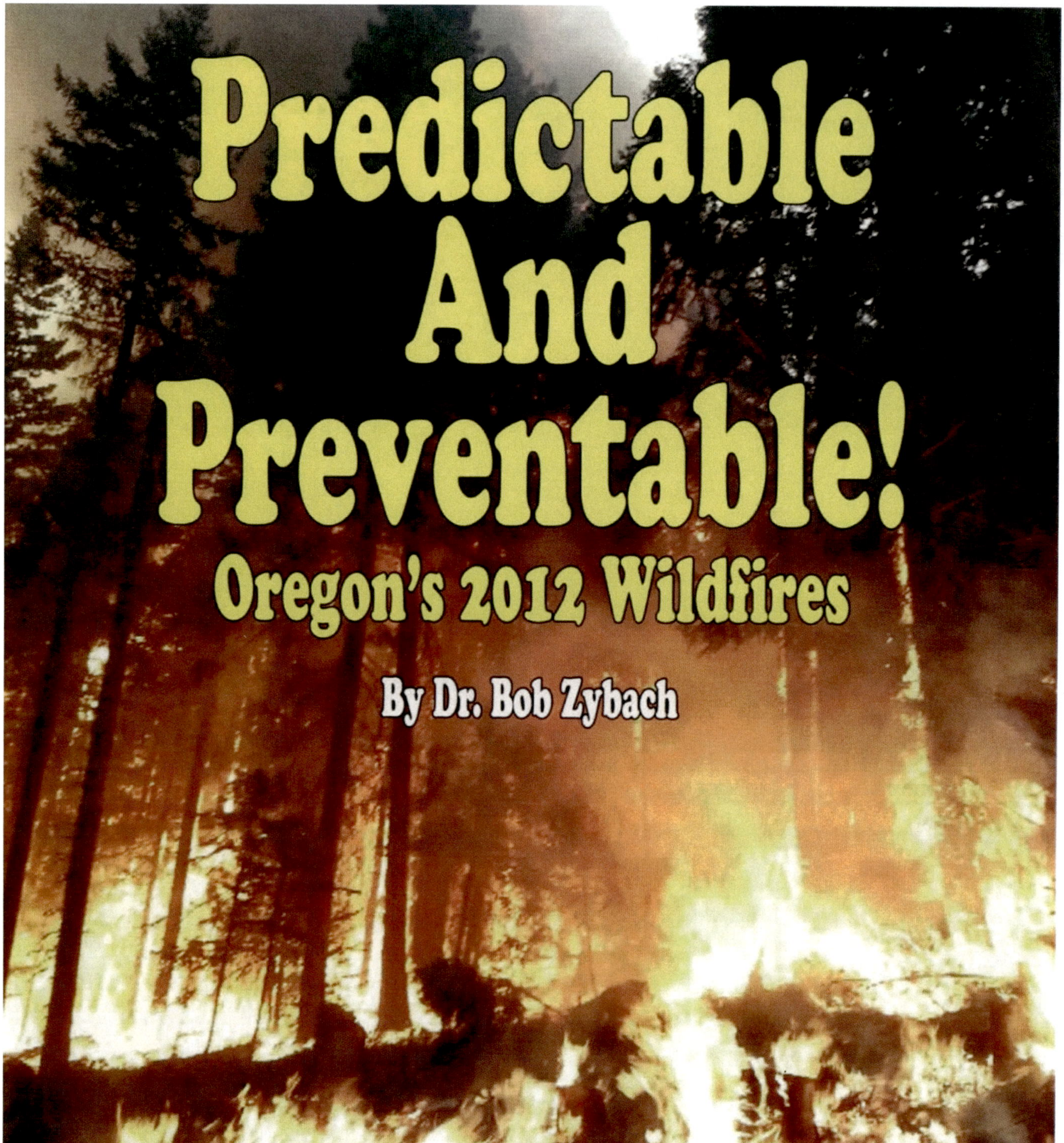

Photo by Gary Sampson Jr. on the August 25, 2012 fire Waterfalls 2 Burnout
http://www.inciweb.org/incident/photograph/3165/3/

This article is the promised follow-up to the Oregon wildfire predictions published in the Summer 2012 issue of Oregon Fish & Wildfire Journal ("Predicting Oregon's 2012 Major Wildfire Events"). It is being written on September 13, 2012, as the current Oregon wildfire season appears to be mostly drawing to an end. However, late September has often been a time of rapid expansion in size of earlier, persistent wildfires and occasionally a time of new wildfire events, such as the Pole Creek Fire that started on September 9 (see Map and Table). Most local hunting seasons begin in early October and have -- when combined with the right weather conditions – also contributed to expansion of existing wildfires and creation of new starts in past years.

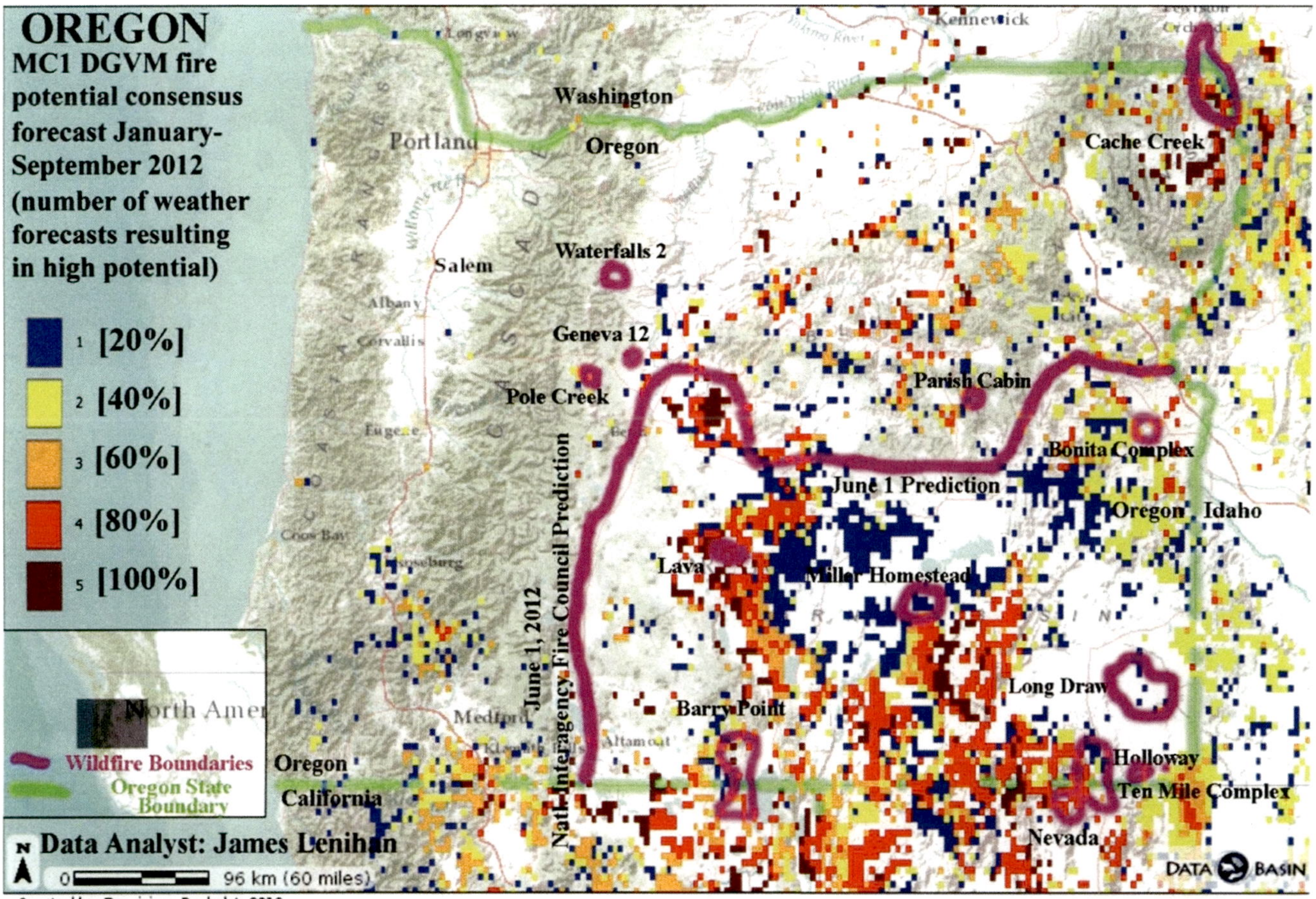

Created by: Dominique Bachelet, 2012

Wildfire Name	Cause	Location	Started	Total Acres
Miller Homestead	Lightning	Burns BLM	July 8	160,853
Long Draw	Lightning	Vale BLM	July 8	557,648
Bonita Complex	Lightning	Vale BLM	July 9	18,188
Lava	Lightning	Lakeview BLM	July 23	21,300
Waterfalls 2	Lightning	Warm Springs Res.	August 4	12,265
Holloway	Lightning	Winnemucca BLM	August 5	461,047
Barry Point	Lightning	Fremont-Winema NF	August 6	93,071
Geneva 12	Lightning	Ochoco NF	August 6	1,337
Ten Mile Complex	Lightning	Vale BLM	August 10	14,036
Cache Creek	Lightning	Wallowa-Whitman NF	August 20	73,697
Parish Cabin	Human	Malheur NF	August 28	6,481
Pole Creek	Unknown	Deschutes NF	September 9	4,583

This year's predictions were based on the historical record of major wildfires in Oregon, beginning in the 1840s; an historical map of lightning-caused fires in Oregon between 1928 and 1932; and two recent predictive maps of this season's events, the first provided by the National Interagency Fire Council on June 1, 2012; and the second by a consortium of US Forest Service, PNW Research Station, and Conservation Biology Institute staff via the Data Basin website (http://databasin.org) on June 8-9, 2012 (also the basis for the current annotated map). Initial indications are that the predictions provided by these disparate sources have been remarkably accurate and generally consistent with one another. The bottom line predictions provided by these combined methods were:

1) Large-scale wildfires, if they occurred, were most likely to take place in eastern Oregon from July through October, and in western Oregon during August and September;

2) The greatest likelihood for the occurrence of major (over 1,000 acres) or catastrophic-scale (more than

100,000 acres) wildfires was the Great Basin region of southeast Oregon -- and with lesser likelihoods of such fires occurring in the central Oregon Bend-Prineville area of the eastern Cascades and Ochoco Mountains, and in the northeast corner of the State, along Hells Canyon;

3) Conversely, the least likelihood of major wildfires was in western Oregon, where risk appeared relatively low and, at best, no greater than "Normal."

The Holloway Fire in Oregon seen from the air on August 11 (Credit: Holloway Fire)
http://www.inciweb.org/incident/photograph/3113/61/

Backside: A view from Holloway Fire burned area Photo by Dave Toney. http://www.inciweb.org/incident/photograph/3113/81/

Fire crew uses hand tools on Holloway Fire in dead grass Photo by Keith Mardis.
http://www.inciweb.org/incident/photograph/3113/22/

So, how accurate were the predictions?

The Table lists 12 major wildfires and their relative sizes – so far -- for the 2012 Oregon wildfire season. This information was obtained from "InciWeb" (http://www.inciweb.org), the federal interagency website created to provide the public and the "Public Affairs community" with a single source of reliable, standardized information regarding all of the nation's wildfire "incidents" during the course of a year. InciWeb is also the source of photographs used in this article to illustrate this year's fires and ground-based fire containment methods.

The Map, an annotated version of the June 8-9 predictive map used in the previous article of this series, shows the relative size and location of each of these fires as of September 13 (Pole Creek has increased in size since then). At this time, according to the InciWeb website, 11 fires are still classified as "Active" -- although most of the updated fire reports have been listing the same stable acreage figures for several weeks or longer. Only the Parish Cabin fire is considered "Inactive" at this time.

The Table and Map also illustrate the precision of this year's predictions – however much "luck" (good and bad) may have had to do with this result: 1) 11 of these fires took place in July or August and one has occurred so far in September, as predicted; 2) the very largest fires have taken place within, or on the perimeter of, the Great Basin, as predicted, and in addition to the 74,000-acre Cache Creek Fire located along Hells Canyon in northeast Oregon -- also as predicted; and 3) there have been relatively few wildfires in western

Stringer of conifers torching, Cache Creek Fire. Photo by Ewert.
http://www.inciweb.org/incident/photograph/3202/9/

Barry Point Fire fuel load burning, August 7. Photo by Fred Way.
http://www.inciweb.org/incident/photograph/3105/23/

Oregon this year, and none of a major size or concern. Again, as predicted. Therefore, this year's major wildfires have all occurred when and where they were predicted to take place -- and have not occurred, so far, in areas in which they were not expected to take place.

In sum, the predictions have been highly accurate, bordering on precise, so far as time and place have been concerned. It is the size of these fires that has been somewhat surprising, with just two of them alone (the Long Branch and the Holloway) totaling more than 1,000,000 acres in size, including a large portion of the latter fire in neighboring Nevada.

When, exactly, did these fires take place?

The largest wildfires in southeast Oregon history began burning on July 8 and 9, ignited by a lightning storm, and on August 6 -10 in the same general area, also started by lightning. In all, four major fires started in July (758,000 acres); eight major fires began in August (662,000 acres); only one major fire so far, Pole Creek, started in September (24,000 acres); and all but one are considered "Active" at this time. This pattern of July to October wildfires in Oregon has been firmly in place for at least 235 years, probably much longer, and seems entirely unrelated to recent claims of Global Warming or "climate change" as affecting wildfire seasonality.

Where, according to ownership, did these fires take place?

Six of these fires began on USDI Bureau of Land Management (BLM) lands, totaling 1,233,000 acres and averaging 205,000+ acres in size; five began on USDA Forest Service (USFS) lands, totaling 199,000 acres and averaging 39,800 acres in size; and one started on the Warm Springs Indian Reservation, totaling more than 12,000 acres in size. All of the July fires started on BLM land, with the most recent BLM fire starting on August 10. The Warm Springs fire started August 4, and all five USFS fires started from August 5 through September 9.

The strong correlation between government land ownership patterns and wildfire starting dates is interesting, and probably related more to fuel types than to management practices. Conversely, the strong correlation between land ownership type (federal government vs. private, state, and tribal) and major wildfire occurrences is probably due more to management practices than to anything else – and which accounts in very large part for the predictability of these events.

Pole Creek Fire: View from Incident Command Post at Sisters Rodeo Ground as temperature inversion blankets Sisters, Oregon in smoke. Photo by Oregon Incident Management Team 4. http://www.inciweb.org/incident/photograph/3244/5/

How severe were these fires?

Wildfires are often described in terms of "intensity" and "severity." Intensity is often described in measurements of fire temperature, flame length, and other considerations of the fuel being consumed, how quickly, under what conditions, and with what immediate results. Typically, this is mostly a consideration of fire heat and the speed and degree in which fuels are consumed; factors dependent on fuel type, ambient temperature, humidity, and local weather conditions at the time and place of the fire.

Severity, on the other hand, is usually little more than a measure of fire-related mortality, particularly in regards to vegetation. Did the fire kill everything in sight, including trees, shrubs, forbs, grasses, and lingering bugs and other wildlife? Very severe. Or did it skip around, sometimes crowning (and killing trees, but maybe sparing understory shrubs and animals) and sometimes skipping from place to place, or burning at relatively low temperatures as it moved along the ground? Moderate severity. Or did the fire mostly just move through an area, consuming dead grasses and some shrubs and bugs in its path, maybe doing more good than harm in those regards? Low severity.

This year's Long Draw Fire, at 558,000 acres, is the largest wildfire, measured in acres, in Oregon history; the 2012 Holloway Fire, at 461,047 acres, is the third largest such fire -- the 2002 Biscuit Fire, at 495,000 acres in southwest Oregon, being the only other historic wildfire of such magnitude as these two. A big difference between these events, however, is the amount of fuel consumed – off-hand I would guess that 10 times as much fuel, maybe more, burned in the Biscuit Fire as burned in the Long Draw and Holloway fires combined. A quick study of the photographs used to illustrate this article reveal why: major portions of this year's fires burned across grasslands and shrublands containing only a few tons of fuel per acre, while the Biscuit Fire – in common with most western Oregon wildfires of similar magnitude – burned through millions of tons of fuels in the forms of massive conifer trees, pitch-soaked snags, large woody debris, and highly flammable shrubs.

Much of the 2002 Biscuit Fire burned at high intensity and resulted in great mortality of plants and animals; many of this year's fires burned at relatively low severity and may actually result in several rejuvenated grasslands and shrub populations -- and the animals that feed upon them and use them as shelter -- as a result. However, it must be pointed out that the majority of lands suffered serious damage and that the costs and problems associated with treating grasslands and shrub populations with prescribed fires instead of waiting for lightning ignitions are significantly safer, cheaper, and less destructive than the risky, chaotic decisions and practices used to manage wildfires.

Summary

The 2012 wildfire season is still in progress as this is being written, but is already among the worst in Oregon history. At this time more than 1,444,000 acres have been burned in the twelve 1,000+ acre fires profiled in this article; the large majority of this acreage is located within the State, but also includes relatively minor amounts in the adjacent states of California (Barry Point Fire), Nevada (Holloway Fire), and Washington (Cache Creek Fire).

These fires, for the most part, were easily predicted. Those predictions were based largely upon historical weather patterns and current fuel patterns. A common observation is that people always talk about the weather, but they never do anything about it -- mostly because we can't, and no matter how hard we may try. Fuel is something else, and it is something that can be (and usually is) managed at one level or another by most people in most times and in most places.

Wildfire fuels in Oregon are generally comprised of pitchy trees (conifers), snags (dead standing trees), shrubs, dead wood, grasses, and litter. The management of these vegetation types, historically, has been primarily achieved via logging, prescribed fire, grazing, pruning, gathering, mowing, irrigation, and other methods typically associated with the growth, maintenance, and harvesting of food, fiber, and fuel products useful to people. The resulting cultural landscapes then form the basis of "wildlife habitat" for the native plants and animals that have best adapted to these practices.

This article documents the predictability of wildfires, using criteria that applies to both Oregon and to much of the remainder of the forests and grasslands west of the Rocky Mountains in North America. Because a major portion of such predictability is based on existing fuel loads, active management of those fuels can mitigate these fires. The first article in this series ("Forest Restoration: Problems & Opportunities," Spring 2012 Oregon Fish & Wildlife Journal) describes how that can be accomplished, to the advantage of America's rural communities and with little or no cost to most taxpayers.

These fires are deadly, dangerous, costly, ugly, unhealthy, and unacceptable. They are also avoidable. Something needs to be done to put an end to this nonsense, in my opinion. This series of articles offers a method and a means for doing so.

Author: Dr. Bob Zybach is a fifth-generation Oregonian and a forester and forest scientist with a long career in the woods of the Pacific Northwest. He has a PhD in Environmental Sciences from OSU and has been Program Manager for Oregon Websites and Watersheds Project, Inc. (www.ORWW.org) since its formation in 1996.

The True Costs of Wildfire & The Douglas Complex

Dr. Bob Zybach and John Marker

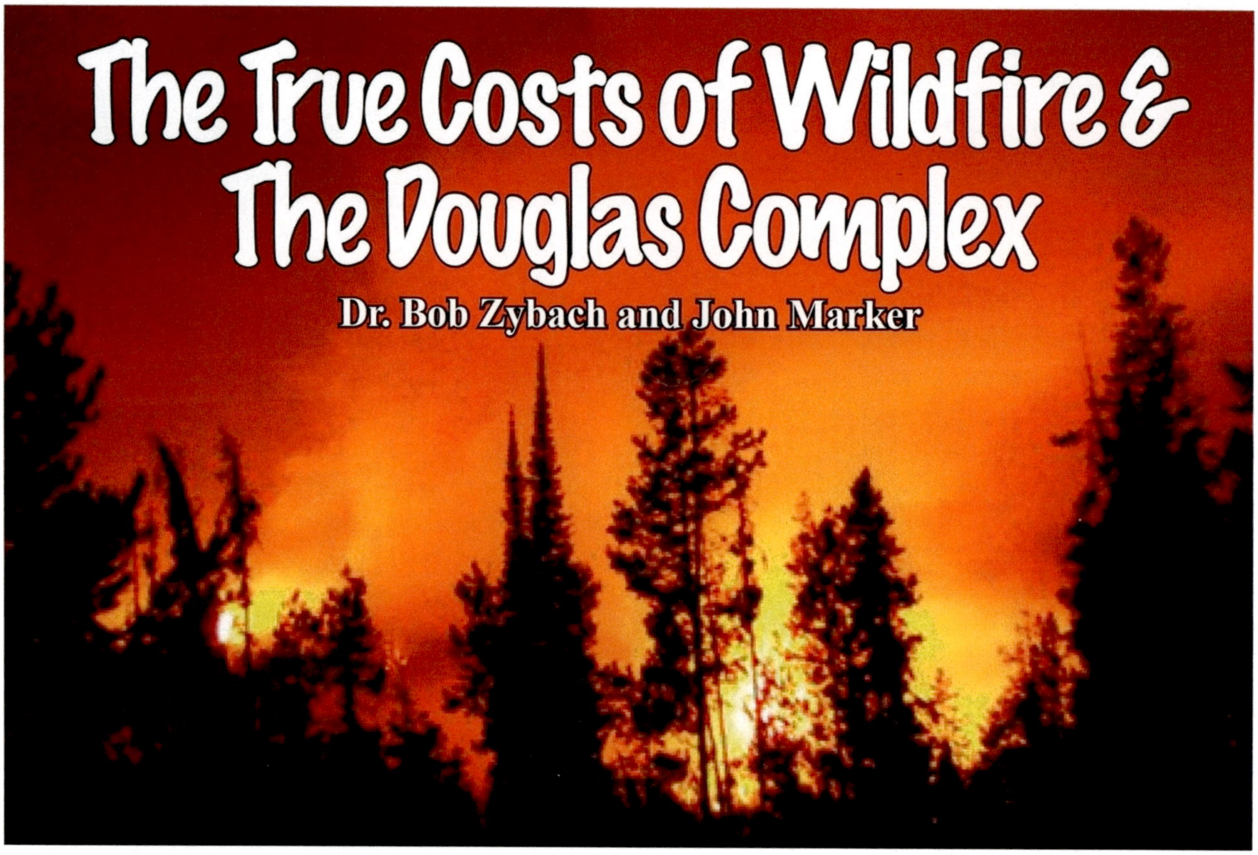

Cascade Complex Fire at night. Picture taken August 8th, 2007, near Warm Lake, Valley County, Idaho. Photo by US Forest Service, provided courtesy of the Yellow Pine Times.

Wildfires have become financial big-ticket items in the United States. The cost of fighting fires continues to escalate, the negative impact on people, wildlife and property grows, and damages to the land and its resources mount. But rarely do we hear discussion of these damages in terms the general public can understand, and when economic damage is discussed it is seldom a front page item. And when wildfire economics are discussed, it is usually in terms of suppression costs and property damage, little else.

As this is being written news has come in today (September 4), that the Douglas Complex fires in SW Oregon have been 100% contained. This Complex was started by lightning strikes on July 26 and was comprised of two large fires, Dad's Creek and Rabbit Mountain, each about 24,000-acres in size.

Three weeks after the fires started, the August 16 Salem Statesman Journal reported: "The Douglas Complex already has burned 46,059 acres and was listed Friday as 65 percent contained. Suppression costs heading into Friday were calculated at $42.25 million, with 2,093 people assigned to it, according to ODF." ODF (Oregon Department of Forestry) is responsible for calculating costs of firefighting for the State's services. But it is the only dollar figure given in the news account, and few readers have any idea what the number actually represents.

On September 4 the Medford Mail Tribune reported: "The 48,679-acre Douglas Complex fire burning just north of Glendale is now 95 percent contained. Total cost of fighting that fire is $51.76 million, ODF reported." This is to illustrate how numbers commonly distributed by local and national media are often limited to basic fire suppression costs – in this instance, still more than $1,000/acre.

Our research has shown that the actual cost of damage caused by the Douglas Complex will likely be much closer to $500 million than to the current figure of $50 million. The fire might be contained, but many of its actual costs and damages are only now beginning to accrue.

The true scope of the problem

The $500 million estimate might sound outrageous to many readers, but examples are easy to come by where it is shown that suppression costs are likely to account for as little as 2% to 10% of the actual damages caused by a large wildfire.

Some examples: In 2009 Western Forestry Leadership Coalition released a report titled, The True Cost of Wildfire in the Western U.S. The authors examined six major US wildfires, and compared suppression costs and tactics with "total costs." Two examples were the 2000 Cerro Grande fire in New Mexico -- suppression costs reflected only 3% of total damage estimates -- and the 2003 Old, Grand Prix, and Padua fire complex in California, in which suppression costs were only 7% of total costs to 2005 – with total losses expected to increase dramatically in years to come (Dunn et al, 2005).

The 2003 fires in San Diego and Southern California were catastrophic by any measure – 24 fatalities, more than 3,700

Home destroyed in 2005 Deer Creek Fire near Selma, Oregon.
Photograph by Bob Zybach, August 26, 2005.

homes destroyed, and suppression efforts were $43 million. However, in 2009, Matt Rahn of San Diego State University presented findings that put suppression costs at less than 2% of the total long-term cost of the fire, www.youtube.com/watch?v=1Y8Ef9qc0F0.

The 2002 Hayman Fire burned 138,000 acres and cost $42 million to suppress. In 2004 Dennis Lynch of Colorado State University estimated that an additional $187.5 million in losses had accrued within a year. Suppression costs were only 18% of the total, and Dr. Lynch stated, "I recognized the need to follow costs into subsequent years to more completely identify a fire's true impact."

On July 12, 2010 National Association of State Foresters released a briefing paper titled State Forestry Agency Perspectives Regarding 2009 Federal Wildfire Policy Implementation. The paper avers that State Foresters have no say so in how the Federal Interagency and Interdepartmental Wildland Fire Management Community fight fires that threaten communities and natural resources. State Foresters would prefer that the Feds implement aggressive fire suppression strategies for any fire with a chance of burning private land and property.

State Foresters recognize that safe and aggressive initial attacks are proven the best suppression response to reduce fire damage and keep costs down – but recognize that the federal agencies are not doing so. They also recognize Federal wildfire management policies impact state fire suppression efforts when Federal fires move across jurisdictional boundaries and burn state protected lands and private property. As such, the federal agencies have increased risk to families, communities, and wildlife by allowing wildfires to burn without containment efforts, and are not providing credible explanations for doing so - not to State Foresters' satisfaction at any rate.

These "let it burn" wildfires are allegedly for resource "benefits" and fire fighter safety, but often blow up, crossing onto state-protected land, putting communities at risk, and placing tremendous burdens on the states to control fires escaping from Federal lands.

This is a serious issue in states and counties where the federal government manages 50, 60 or even 90 percent of the land, and pays no land taxes. Further complicating the situation is an apparently influential group of people claiming wildfire is "natural" and the land will "heal" and everything will be better if nature is allowed to "take her course" -- they do not explain, however, why such "ecosystem services" can't be provided more effectively, safer, and with far less risk, fear and cost to taxpayers and wildlife via prescribed fires, ignited by people with stated objectives and written plans.

By mid August of this year the federal government had spent more than $1 billion fighting wildfires in the West, creating an estimated $10 billion to $25 billion in actual costs and damages. During the same time states and local fire units have spent hundreds of millions more for fire fighting. On August 21, 51 wildfires of significant size were uncontrolled.

Pacific Crest Trail, B&B Complex Fire. Photograph by Bob Zybach, May 2004.

On August 22 the US Forest Service announced it was going to take more than $600 million from non-fire programs to pay its anticipated 2013 fire fighting costs.

What Can Be Done?

There is little discussion of local economic damages caused by these fires other than structural losses, and often such losses are reported without mentioning the dollar amounts involved. However, the federal agencies are constantly challenged by members of Congress to reduce fire fighting costs, but without any sense of total fire costs.

Four years ago we were faced with the same problems and questions, and it appears little has changed: these problems took quite a while to become established, and they will take longer before they can be resolved. At that time a small group of individuals with similar backgrounds, interests and concerns in these matters asked ourselves what we could do to help resolve these problems. It was easy enough to come up with examples and complaints, but it was difficult to figure out how to make things better – particularly for such a small group.

The result of these discussions was an informal ad hoc – and truly grassroots – effort we titled the "Wildfire Cost-Plus-Loss Project." Our intent was to develop analytical tools and sources of information that could be used by most citizens and not limited to agencies, professional organizations, or special interest groups.

The focus of much of our efforts was to design a simple tool that could be used effectively by almost anyone to assess the true damages of large wildfires. Target audiences were students, journalists, landowners, residents, elected officials, insurance adjustors and resource managers. Our initial efforts resulted in: 1) development of a "one pager" wildfire damage checklist/accounting form that could be used by almost anyone with access to the news; 2) a peer-reviewed article published online with an appropriate federal agency, including instructions for using the one-pager;

Metolius River Basin, B&B Complex Fire. Photograph by Bob Zybach, May 2004.

and 3) a public informational website for anyone to use who was interested in the topic or use of these tools.

One-pager worksheet: www.wildfire-economics.org/Checklist/One_Pager_Checklist_2009.pdf

Peer-reviewed online publication with a federal agency: www.wildfirelessons.net/Additional.aspx?Page=240

Informational website (under construction): www.wildfire-economics.org/index.html

These efforts were generally well received and promoted on local, statewide, and national media during several high-level meetings, but have never been properly field tested or adopted. In our opinion, not considering the true economic impact of wildfires to governments, businesses, and people is an unfortunate omission. For resource managers it is difficult to explain the realities of fire protection and its nuances to Congress and to citizens directly affected. The lack of communication also implies that the land has no economic value for production of resources or public enjoyment. To the public the impression can be that it is government waste spending money fighting fire if there are no damages.

The "one-pager"

The focus of much of our group's efforts was to design a simple tool to could be used effectively by almost anyone. Target audiences were students, journalists, landowners, residents, elected officials, insurance adjustors and resource managers.

Table 1. Wildfire 'Cost-Plus-Loss' Ledger Checklist Form

Fire Name __________________ County ________ State _____ Country __________
Ignition Date ____________ Containment Date ____________ Total Acres __________
Cause: Human___, Lightning__, Operation__, Prescription__, Maintenance__, Other___
Major Landowner(s) __________________ Human Fatalities ____ Homes Lost _______

Cost-Plus-Loss Category	A. Direct	B. Indirect	C. Post Fire	Totals
1. Suppression Costs				
a. Public				
b. Tribal/Private				
2. Property Damage				
a. Public				
b. Tribal/Private				
3. Health Effects				
a. Public				
b. Tribal/Private				
4. Vegetation				
a. Public				
b. Tribal/Private				
5. Wildlife				
a. Public				
b. Tribal/Private				
6. Water				
a. Public				
b. Tribal/Private				
7. Air and Atmospheric				
a. Public				
b. Tribal/Private				
8. Soil-Related				
a. Public				
b. Tribal/Private				
9. Recreation				
a. Public				
b. Tribal/Private				
10. Energy				
a. Public				
b. Tribal/Private				
11. Heritage				
a. Public				
b. Tribal/Private				
Totals				

The one-page checklist is intended to make initial estimates – based entirely on available data and personal estimates -- of total fire costs, and to ultimately be used in conjunction with a comprehensive ledger for better tracking costs and losses over time. We believe the use of these tools would better inform land and resource managers in the management of water, fuels and wildfires by identifying true costs of decisions and allowing better judgment in the establishment of resource use priorities. Such uses would also generalize discussion topics and terminology for better communication with the public.

The checklist is divided into 11 categories: Suppression costs; Property; Public health; Vegetation; Wildlife; Water; Air and atmospheric effects; Soil-related effects; Recreation and aesthetics; Energy; and Heritage (cultural and historical resources). Each of the categories was considered in terms of direct costs (e.g., fire suppression, lost lives, evacuations, burned homes, etc.), concurrent indirect costs (fire preparedness equipment and training, fire insurance premiums, air and water quality, aesthetics, etc.), and post-fire costs (long term damages to society and the environment; e.g. loss of timber, crops, and wildlife habitat, chronic human health problems, reservoir sedimentation, etc.).

By using this basic approach we think the economic effects of large wildfires can be readily quantified, compared and described in terms understandable to lawmakers and to the public. By combining this data with digital spreadsheets, we think better ecological, economic and strategic decisions can be made in the management of our common lands and resources.

Conclusions

Few people seem to understand the critical importance of our nation's natural resources to the future of the United States. Calculating economic damages may not be the best way to describe the total impact of wildfire on the land and people, but it is a method of creating awareness of damage by using a vehicle that most people understand: money. An analysis of actual wildfire damages provides needed context for evaluating protection and prevention programs, as well as overall natural resource management goals and objectives.

While we have focused on economics we are very much aware of biological considerations, their complexities, their importance to meeting the natural resource needs of 300 million people -- but that's for another study. Now is a better time to test our earlier efforts and conclusions, and perhaps the Douglas Complex provides and ideal circumstance for doing so.

Global Warming and Oregon Wildfire History

Interview with Dr. Bob Zybach

From the Lars Larson Radio Show, July 29, 2014

Transcribed by Nana Lapham

Human interference — such as politics and modern techniques to protect forests from blazes — allows tree parasites to lay waste to 90,000 acres of the Deschutes and Willamette national forests.

A natural disaster runs amok

By Theresa Novak
The Statesman Journal

ENVIRONMENT

Tiny worms and voracious beetles that have decimated almost 90,000 acres of trees are almost gone. Many of the dead trees they left behind soon will be gone, too.

For the first time in almost four years, the U.S. Forest Service will sell salvage timber sales on the Sisters and Mckenzie ranger districts. There, drought and infestation by spruce budworms and bark beetles have killed a vast number of trees in an area bisected by the Santiam Pass.

The damaged area crosses over two federal forests and encompasses some private timberland. Most of the damage is in the Deschutes National Forest, with a smaller portion in the Willamette National Forest near Suttle Lake.

Rich Mathis, who has owned the Suttle Lake resort for three years, said it is about time that the Forest Service did some-

September 3, 1994 Salem Statesman Journal article describing 90,000 acres of beetle killed US Forest Service land along Highway 20 and the Santiam Pass.

In the summer 2013 issue of Oregon Fish & Wildlife Journal I wrote an article about the "science" behind spotted owl listings and current federal forest management policies. In late July, about a month after the article first appeared, Lars Larson interviewed me about its content on his popular radio show. Almost exactly one year later -- to the day and hour -- I was on Lars' radio show again, this time regarding recent news statements by "top government officials" regarding forest fires and global warming.

I have been interviewed by Lars five or six times over the course of the past 12 or so years, beginning sometime in 2002 with an interview that was also likely about forest fires; perhaps the Biscuit Fire of that year. Recently, two members of Lars'staff, Tony King and Brandon Christensen, located and sent me audio MP3 recordings of my two most recent interviews. My long-time research associate at www.

ORWW.org, Nana Lapham, then transcribed both so they could be made available online.

So now we are making an article from a Lars Larson interview, rather than being interviewed about an article. The following transcript is significantly shortened, mostly to the advantage of my own statements and the detriment of insightful comments and observations Lars is making regarding climate change and forest management. The full interview, in both aural MP3 files and PDF transcriptions, was mentioning that he had visited the High Desert Museum and he was interviewed by KTVZ, just so we get all the credit in and he said because of climate change and biomass, wildfires are different now than they were ten years ago. And a direct quote from Tidwell, 'Our fire seasons today are 60 to 80 days longer, which creates more energy in the system.' Tidwell told News Channel 21, 'We see more erratic fire behavior.' Let's start with the length of the fire season and I want to ask you first. You did your PhD thesis on this

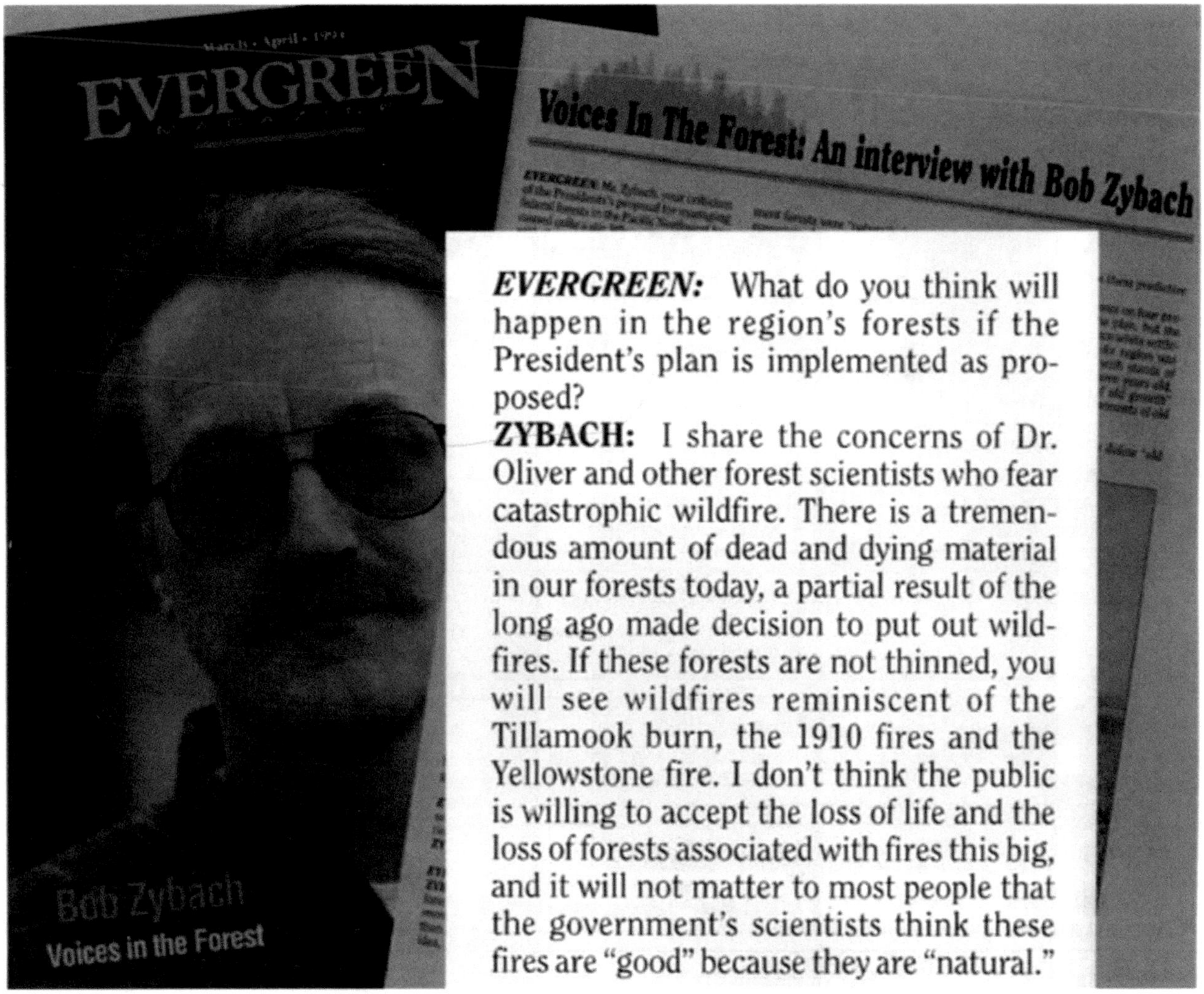

EVERGREEN: What do you think will happen in the region's forests if the President's plan is implemented as proposed?

ZYBACH: I share the concerns of Dr. Oliver and other forest scientists who fear catastrophic wildfire. There is a tremendous amount of dead and dying material in our forests today, a partial result of the long ago made decision to put out wildfires. If these forests are not thinned, you will see wildfires reminiscent of the Tillamook burn, the 1910 fires and the Yellowstone fire. I don't think the public is willing to accept the loss of life and the loss of forests associated with fires this big, and it will not matter to most people that the government's scientists think these fires are "good" because they are "natural."

March-April 1994 Evergreen Magazine interview with predictions of catastrophic wildfire in untended forests in the western US, but specifically referencing the Santiam Pass and the Silver Complex burn areas that became the B&B Complex and the Biscuit Fire - each the largest in history for their region.

can be found at www.NWMapsCo.com

LL: Welcome back to the Lars Larson Show. You know that we talk about global warming issues on this show on a pretty regular basis and we love to have people on both sides of the question on … Bob, good to have you back on and thank you very much for taking the time today.

BZ: Well thank you, Lars. I enjoy it.

LL: Tom Tidwell is the U.S. Forest Service Chief and it subject, right?

BZ: Yes, on the history of Coast Range wildfires, but . . . the last 20 or 30 years, I've looked very closely at the history of wildfires in the Pacific Northwest.

LL: Are the fire seasons two to two and a half months longer than they used to be?

BZ: No, they are about exactly the same for the last 200 years.

LL: [Laughs] That's quite a statement. So for 200 years, we've had fire seasons about the same length. Why do you suppose the U.S. Forest Service Chief, Tom Tidwell, is saying then that they are two to two and a half months longer?

BZ: Well maybe because he is following lead of President Obama, who said last week up in Washington that the

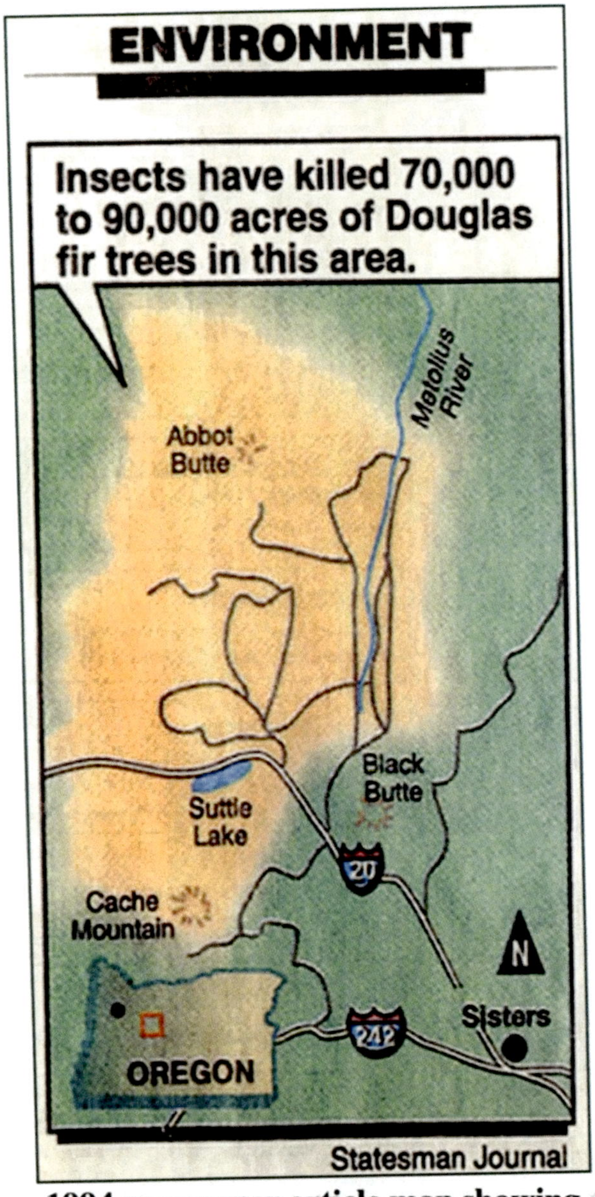

1994 newspaper article map showing 90,000 acres of dead trees on the Santiam Pass. Note the locations of the same landmarks. Map color choice was complete coincidence.

wildfires are due to "climate change. " I think they have agendas to support. I think it's political. It's not scientific and that's what bothers me.

LL: They say outside of the west coast . . . burn acreage so far this year in the United States is the lowest in a decade, less than half of normal and one of the lowest on record.

BZ: Well I don't know if it is one of the lowest on record, and I don't know how they are measuring "normal," but the last ten years have been terrible and the last 20 years have been terrible. Actually going back to about 1987. But these fires were predicted more than 20 years ago by Chad Oliver up at University of Washington (he's at Yale now), by myself, by Jim Peterson, who has Evergreen Magazine -- which was nationally distributed -- that the build-up of fuels caused by passive management of our forests was going to result in massive forest fires and that's exactly what has occurred.

LL: . . . Is it warmer today than it was ten years ago or

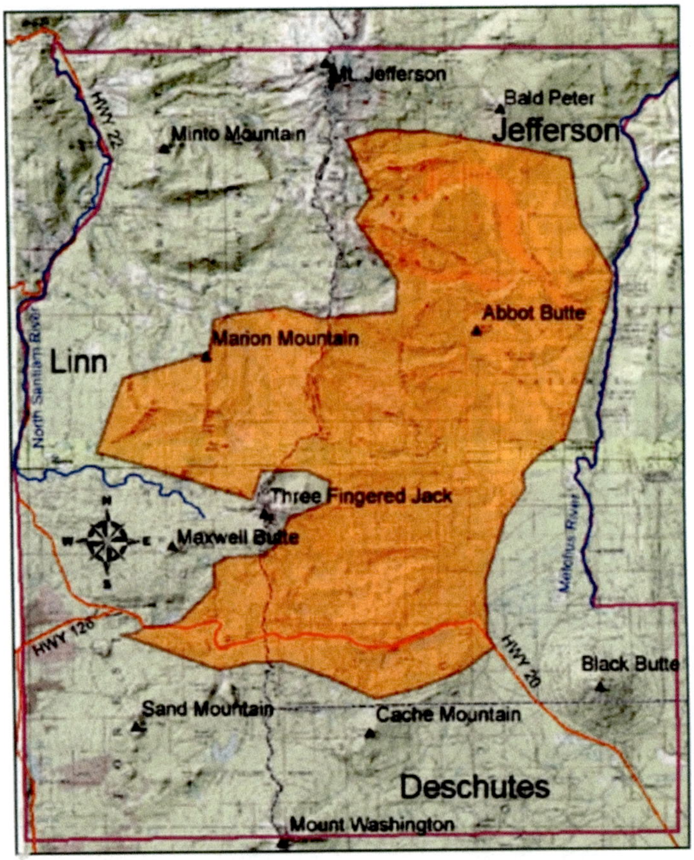

2004 ORWW website report on the B&B Complex showing 90,000 acre fire boundaries. Note the locations of the same landmarks. Map color choice was complete coincidence.

cooler or about the same temperature?

BZ: I think the records are showing that in the northwest we've got generally cooling temperatures for the last 15 or 17 years. As far as the seasonality of wildfires, they're pretty darn predictable. July and August are the big months. Have been since the 1820s. Sometimes it will start a little bit earlier in June or early July in eastern Oregon, because it dries out sooner. But western Oregon, about now [late July] until mid-August is when it gets its worst and then they usually taper off in September. There's some starts there, and then the rains hit them and that's been the same pattern as long as we've had historical records. It has not changed.

LL: . . . When Tidwell says something like that and says the fire seasons are longer today than they used to be . . . You know, small, relatively small changes in temperature are not going to make the difference between say the Coast Range where it's wet and damp and central Oregon where it's bone dry all summer, is it?

BZ: Not even ten or 20 degrees. We had the wildfires here in February and March. It depends on humidity and the wind. You get an east wind and low humidity and the fuels dry out. So fuel build-ups are a result, in Oregon, of spring rains. That's where you get your vegetation build up: or management policies that say don't salvage dead material; or let things grow and don't log, say, in forests; or don't graze on the prairies. So it's not a matter of temperature at all. It's humidity and wind and it's fuel. And then it's a source of ignition; and since we have lightning strikes -- not on the Coast Range, but southwest Oregon, eastern Oregon and along the Cascades -- at this time of year, it's predictable that when they strike fuels, they will burn. If we have an east wind and low humidity they get out of control really quickly. But the temperature doesn't have anything to do with the length of fire season and the main thing that controls the annual fuels, the ones that are particularly combustible, are the spring rains, so you get big heavy grass build ups and a lot of [woody] growth.

Fall 2003 cover of Oregon Fish & Wildlife Journal. Showing sign posted in 1994 and published in the Fall 1994 Issue, warning drivers of beetle kill area to the point of trees potentially falling.

LL: On the subject of fire seasons, so we have the head of the U.S. Forest Service saying that we have a longer fire season by 60 to 80 days . . . Let's talk about the fuel load . . . Give me your take on that first, if you don't mind, Doc?

BZ: Sure. This probably started -- the fuel load problem that has led into the wildfires -- probably started with the Wilderness Act. A lot of Wilderness areas included timber. The Kalmiopsis in southern Oregon is a good example. In 1987 it burned, there was no salvage because it was a Wilderness area. The Jefferson Wilderness area had beetle-kill in the early 90s. In both those instances, myself and other people said these are going to burn up catastrophically if something isn't done to remove these now dry fuels. Anybody that has used wood heat knows that dry wood burns better than green wood. And then point of fact, they both did. The B&B Complex burned up almost the exact perimeter of the beetle-kill on the crest of the Cascades in [2003]. The [previous] year, the Biscuit fire burned up the Kalmiopsis again, took out the entire Silver Complex fire that had burned there in 1987, and expanded the boundaries to about 500 thousand acres. So those things were predicted before we had a global warming problem, or about the same time. People, myself included, were saying these are time bombs; these fuels are going to go out of control. Since then, we've exacerbated the problem by two things: [1] we've created a wildlife habitat for spotted owls, which is burning up for the same reason -- the fuels are building up and there's nothing to control them; and [2] the other reason is lack of salvage. We are not salvaging the areas that have been killed and so we are piling up the dry firewood in those areas and as a result the fires are, as predicted 20 years ago, becoming larger and more destructive.

LL: And when they talk about trying to ameliorate this through thinning projects . . . That doesn't actually solve the problem, or probably even reduce it, does it?

BZ: Not a bit . . . I've seen a lot of those projects and they are pretty irritating. They are just busy work. Before I went back to college, I worked for 20 years in reforestation, so we did lots of reforestation work throughout the Pacific Northwest and by removing fuels, by putting in road systems, by using aerial surveillance and so on, we were able to keep wildfires at bay from the end of World War II say, until the 1980s and until the Federal government started putting in passive management policies. The thinning projects they put in are silly, almost. They encourage the remaining trees to grow [greater amounts of fuel]. They are more like precommercial thins or leaf-raking, almost literally, than they are anything that needs to be done. We need to go in and be very aggressive and actively manage the major fuels, which are logs, which also create jobs, which also create income, which also create safer conditions for fish and wildlife and less air pollution and less water pollution.

LL: And if you do that, and you, you know in the process of logging, you are knocking off the limbs, you take the limbs and grind them up, I don't know how much commercial value that grind is, but there are markets for some of that stuff and if you take the rest of it and burn it . . . you should be able to do all that and actually improve the forest, rather than just let it build up, until it's feet thick and then a fire starts and the fire sits there and burns through this fuel from now till the rains hit.

BZ: And chars the soils and sterilizes it and it sloughs off in the rivers and the whole bit; and creates a lot of carbon dioxide in the process for the people that are worried about that. The way the Indians did it for ten thousand years or so is to have regular prescribed burns and so the fuels never built up and the nutrients that go into the soil or the ash from the plants that [burned] – the flash fuels mostly -- and those are the forests that are uniformly described as Eden-like by the first explorers and settlers that came to Oregon. The forests were safe. They were beautiful. They were highly productive. Wildlife was productive. So a combination of logging to get rid of all the fuel build ups of the last 30 and 50 years -- but done along the watershed boundaries and on a landscape scale -- combined with prescribed fires as the Indians did, would pretty much resolve the problem as well as rejuvenating our rural economies and our wildlife populations.

"... we've exacerbated the problem by two things: [1] we've created a wildlife habitat for spotted owls, which is burning up for the same reason the fuels are building up and there's nothing to control them; and [2] the other reason is lack of salvage. We are not salvaging the areas that have been killed and so we are piling up the dry firewood in those areas and as a result the fires are, as predicted 20 years ago, becoming larger and more destructive."

LL: . . . Dr. Zybach, I appreciate you coming on. We are going to have some links up to some of the material that you have and you always bring a lot of intelligence to the show. Thanks very much.

BZ: Thank you Lars.

LL: You bet you. Dr. Bob Zybach with me. You've got the Lars Larson Show.

Historic Wildfires of Western Oregon, 1765 to 2014

By Dr. Bob Zybach

In 1934 William G. Morris, a forest scientist with the USDA Pacific Northwest Forest Research Station, wrote an article for the Oregon Historical Quarterly titled "Forest Fires in Western Oregon and Western Washington." A few months ago Jim Rombach, retired Weyerhaeuser forester and principal consultant of Rombach & Associates, commissioned me to update Morris' work. This article presents my current findings regarding the Oregon portion of that project.

A lot of major fires have taken place in western Oregon since 1934 (mostly beginning in 1987), and a lot more has been learned about the early fire history of the region since that time through scientific research and computerized mapping and analysis. To illustrate the "too incomplete" lack of historical data claimed by Morris, he even footnoted his article with a plea "to any reader having knowledge of additional facts or sources of information" to contact him directly at his address at the US Courthouse in Portland.

Morris was also the author of a definitive report on the 1933 Tillamook Fire and another on the history of lightning strikes and fires in Oregon. His work was influential in the writings and careers of such writers and foresters of that time as Stewart Holbrook – who wrote a well-known book in 1945 on catastrophic wildfires of the US (including the Tillamook Fires of 1933 and 1939) – and leading industrial forester Bill Hagenstein. These men, and others, helped form the Keep Washington Green and Keep Oregon Green movements that became instrumental in bringing public awareness to this problem.

As a partial result of these efforts, there were no major forest fires in western Oregon – excepting the Oxbow Fire of 1966 – in the 26-year period from 1951 until 1987. Since then there have been numerous such fires, mostly ignited by lightning and arsonists and mostly taking place on federal lands in on USDA National Forests (created by Congress in 1905), and on USDI BLM O&C Lands (created in 1937), and USDA Wilderness areas (created in 1964).

The following table depicts the major wildfires that have taken place in western Oregon from ca. 1765 until the present. These fires were not limited to forestlands, but that is where the large majority occurred. Criteria were fire size (typically,

Historic Wildfires of Western Oregon

Wildfire Name	Year	Mo.	Acres	County 1	Notes & Agencies	Cause
Millicoma	1765	?	200,000	Coos		Unknown
Mt. Hood	1800	?	10,000+	Multnomah	Volcanic eruption	Volcano
Willamette Valley	1826	9	10,000+	Yamhill	Indian burning fires	People
Wilkes Expedition	1841	9	1,000+	Jackson	Indian burning fires	People
Barlow Road	1845	10	1,000+	Clackamas	Wagon road-clearing fire	People
Oregon City	1848	7	1,000+	Clackamas	Provisional Government	Unknown
Yaquina I	1849	8	450,000	Lincoln	Indian and settler fires	People
Nestucca	1853	0	350,000	Tillamook	Indian and settler fires	People
Canyon Creek	1856	?	10,000+	Linn	US Territory	Unknown
Columbia Complex	1857	8	10.000+	All	Indian and settler fires	People
Jacksonville Complex	1864	8	10,000+	Jackson	Indian and settler fires	People
Silverton	1865	8	100,000+	Marion	Ranching and logging lands	People
Seven Mile Hill	1867	9	20,000+	Linn	Wagon road-clearing fires	People
Yaquina II	1868	8	300,000	Lincoln	Indian and settler fires	People
1868 Complexes	1868	8	200,000+	All	Indian and settler fires	People

Historic Wildfires of Western Oregon 1765-2014

Wildfire Name	Year	Mo.	Acres	County 1	Notes & Agencies	Cause
Coos	1868	9	125,000	Coos	Indian and settler fires	People
Silver Creek Falls	1886	8	1,000+	Marion	Ranching and logging lands	People
Barlow Road Complex	1901	8	85,700	Clackamas	USDI Cascade Reserve	People
1902 Complexes	1902	9	200,000+	All	27 deaths; 5+ towns	People
1910 Complexes	1910	8	100,000+	All	1 death; USDA NFs	People
Cedar Butte	1918	9	40,000	Tillamook	Ranching and logging lands	People
Astoria	1922	12	N/A	Clatsop	2 deaths; 30 blocks	People
Tillamook I	1933	8	311,000	Tillamook	1 death; ODF; CCC	People
Bandon	1936	9	N/A	Coos	11 deaths; 500 buildings	People
1936 Complexes	1936	9	50,000+	All	Ranching and logging lands	People
Tillamook II	1939	8	209,700	Tillamook	ODF; CCC; industrial lands	People
Tillamook III	1945	7	182,000	Tillamook	3 deaths; ODF	People
Hubbard Creek	1951	8	15,600	Curry	Industrial forestlands	People
Tillamook IV	1951	4	32,700	Tillamook	Fern burners; ODF	People
Roseburg Blast	1959	8	N/A	Douglas	14 deaths; 300 buildings	People
Oxbow	1966	8	42,900	Lane	1 death; USDI BLM	People
Silver Complex	1987	8	96,000	Josephine	Kalmiopsis Wilderness	Lightning
Bland Mountain	1987	7	10,300	Douglas	2 deaths; 8 homes; private	Arson
Warner Creek	1991	10	9,000	Lane	USDA Willamette NF	Arson
East Evans	1992	8	10,100	Jackson	USDI BLM O&C Lands	People
Hull Mountain	1994	8	8,000	Jackson	I death; 44 buildings	Arson
Charlton Fire	1996	8	10,400	Lane	Waldo Lake Wilderness	Lightning
Apple	2002	8	17,600	Douglas	USDA Umpqua NF	Arson
Biscuit	2002	7	500,000	Josephine	Kalmiopsis Wilderness	Lightning
Tiller Complex	2002	7	69,800	Douglas	USDA Umpqua NF	Lightning

Historic Wildfires of Western Oregon 1765-2014

Timbered Rock	2002	7	27,400	Jackson	USDI BLM O&C Lands	Lightning
B&B Complex	2003	8	90,800	Linn	Jefferson Wilderness	Arson?
Deer Creek	2005	8	1,548	Josephine	8 homes; USDI BLM	People
Blossom	2005	7	14,800	Curry	Wild Rogue Wilderness	Lightning
Middle Fork	2008	8	21,100	Jackson	Sky Lakes Wilderness	Lightning
Rattle	2008	8	19,800	Douglas	USDA Umpqua NF	Lightning
Boze	2009	9	10,600	Douglas	USDA Umpqua NF	Lightning
Rainbow	2009	9	6,100	Douglas	USDA Umpqua NF	Lightning
Tumblebug	2009	9	14,600	Lane	USDA Willamette NF	Lightning
Oak Flat	2010	8	7,500	Josephine	USDA Rogue River NF	People
Dollar Lake	2011	8	6,300	Clackamas	Mount Hood Wilderness	Lightning
Douglas Complex	2013	7	48,700	Douglas	USDI BLM O&C Lands	Lightning

only fires in excess of 10,000 acres have been listed); human mortalities; and physical damage to human structures, such as homes, barns, office buildings, powerlines and the like.

A close examination of this table reveals several interesting patterns that should be of interest to western Oregon resource managers and US taxpayers - and to their elected representatives:

1) The western Oregon wildfire season has been about the same for the past 175 years — about 45-60 days long, generally beginning in late July or August and peaking in August or September, and ending with fall rains in September or October. Despite recent public claims of President Obama, Washington Governor Inslee, and US Forest Service Chief Tidwell in Oregon and Washington this past summer, fire seasons are not getting longer and "Global Warming" apparently has nothing to do with it.

2) Almost all major fires dating to precontact and early historical times were caused by Indians or American settlers. The first lightning-caused fires, for example, were not reported in the Siuslaw National Forest until 1927 and they did not reach historic proportions. This is despite the regular and predictable timing and locations of lightning strikes

This map of Oregon lightning-caused fire patterns based on 5300 fires reported from 1925 to 1931 was included in William Morris' June 1934 report, *Lightning Storms and Fires on the National Forests of Oregon and Washington.*

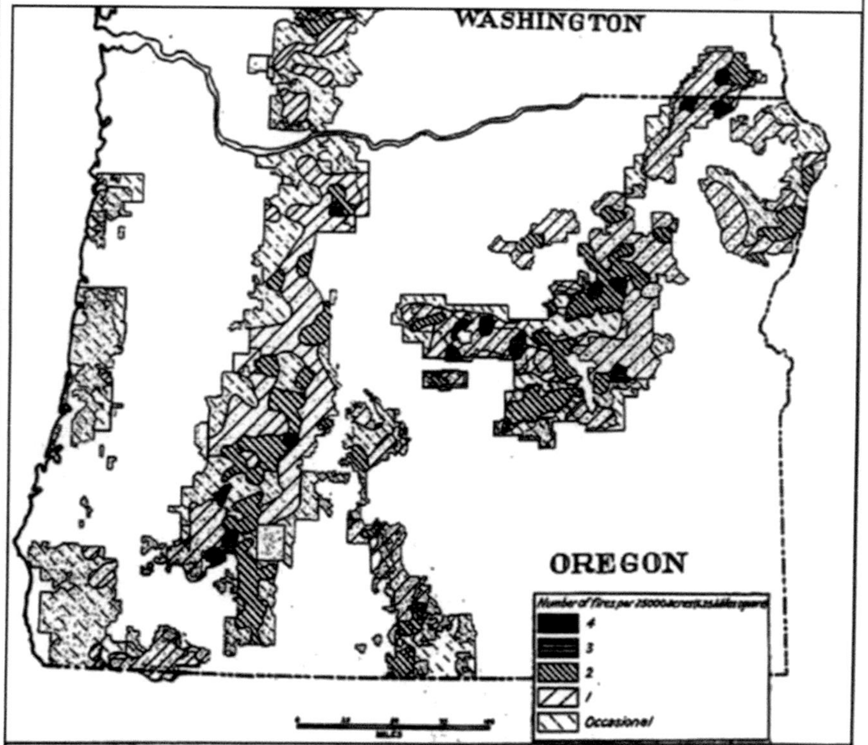

FIGURE 14.--Zones of average yearly lightning fire distribution on the national forests of Oregon and Washington obtained by plotting the locations of the 5300 lightning fires reported from 1925 to 1931.

throughout the region during historical time; most of which are confined to southwest Oregon and the western Cascades.

3) Each county has its own history of large-scale wildfires, with significant differences between them: e.g., Tillamook County had numerous such fires from 1853 until 1951, and little or nothing since; while Douglas County had few major fires until 1987, and have seemingly had them on an almost annual basis since.

4) There were hardly any major wildfires in western Oregon between 1951 and 1987; a period in which these forests were the most actively and intensively managed in history, and during which time the "Keep Oregon Green" movement was in full swing.

5) Almost all major wildfires during the 28 years since 1986 have occurred on federal lands and were ignited by lightning or arsonists; further, these fires were predicted by several scientists and other knowledgeable professionals more than 20 years ago due to major changes in federal land and fire management policies and had little or nothing to do with "fire suppression history" or "climate change," as has been suggested.

In sum, predictable large-scale wildfires are largely preventable, as evidenced by the success of the Keep Oregon Green movement and by the seasonal fire control efforts of the Oregon Department of Forestry in curbing most such fires on state and private lands in the 70 years since World War II. Conversely, passive forest and fire management actions and policies -- such as those associated with Wilderness areas, spotted owl habitat, and "let it burn" wildfires – that were implemented on federal lands beginning in the late 1960s seem to be primarily responsible for the severity and extent of most forest fires in western Oregon during the past 28 years. As predicted.

Dr. Zybach has a PhD in Environmental Sciences from Oregon State University. His dissertation research was titled The Great Fires: Indian Burning and Catastrophic Forest Fire Patterns of the Oregon Coast Range, 1491-1951.

B. THE GREAT YAQUINA BURN.

U. S. GEOLOGICAL SURVEY

These photos of the "Great" Nestucca and Yaquina burns that resulted from the 1849-1868 Yaquina and Nestucca fires appear in Henry Gannett's 1902 report to Congress, *The Forests of Oregon*

The author is shown discussing historical maps at the Round Lake location of the B&B Complex fire on September 15, 2004. Also pictured are (clockwise from foreground) Kermit Cromack, Jim Peterson, Benjamin Stout, Zybach, Wayne Giesy, and Nana Lapham (photo by Brett Morrisette, USDA Pacific Northwest Research Station).

Eisenhower Was Right!

Government Science is Killing Our Forests

By Bob Zybach PhD

Excerpt from President Dwight D. Eisenhower's Farewell Address, January 17, 1961

"...Akin to, and largely responsible for the sweeping changes in our industrial-military posture, has been the technological revolution during recent decades.

"In this revolution, research has become central; it also becomes more formalized, complex, and costly. A steadily increasing share is conducted for, by, or at the direction of, the Federal government.

"Today, the solitary inventor, tinkering in his shop, has been overshadowed by task forces of scientists in laboratories and testing fields. In the same fashion, the free university, historically the fountainhead of free ideas and scientific discovery, has experienced a revolution in the conduct of research. Partly because of the huge costs involved, a government contract becomes virtually a substitute for intellectual curiosity. For every old blackboard there are now hundreds of new electronic computers.

"The prospect of domination of the nation's scholars by Federal employment, project allocations, and the power of money is ever present and is gravely to be regarded.

"Yet, in holding scientific research and discovery in respect, as we should, we must also be alert to the equal and opposite danger that public policy could itself become the captive of a scientific technological elite."

During the past five years I have written a number of articles and editorials for this magazine that have specifically looked at federal forest management policies, laws, and regulations and the so-called "best available science" that is said to be their foundation. These writings were mostly inspired by the massive changes that have taken place on federal forestlands in Oregon during my lifetime and that have directly resulted in ruined rural economies, broken families, depleted wildlife populations, air and water pollution, degraded landscapes, and hundreds of thousands of acres of dead and dying trees. Almost all of it unnecessary and preventable.

Most of the articles focused on specific topics, such as wildfire economics, spotted owl habitat, streamside buffer regulations, cattle grazing along fish bearing streams, and catastrophic wildfire mitigation. This article is essentially a "bottom line" summary of these earlier writings, and less detailed as one result.

The title to this article/editorial is somewhat misleading in order to be concise and provocative. Eisenhower didn't actually say that "government science" might result in the adverse descriptions given above; he said that government funded scientific research could compromise "intellectual curiosity" and potentially result in misguided policies dictated by a "scientific technological elite" (see excerpt). That is the very process that most concerns me and is typified -- at least in my mind -- by the catastrophic wildfires that have been ravaging our federal forestlands and rural counties the past 30 years.

1961 Eisenhower Speech

I was a 12-year-old 6th-Grader in Baker, Oregon when President Dwight D. Eisenhower delivered his famous "Military-Industrial Complex" speech on January 17, 1961. This was his televised farewell address to the American people and, even though it has been widely-viewed and quoted ever since, I never read a complete transcript or watched the entire speech until a few years ago.

When I read the complete speech and watched it on YouTube, I was shocked at the accuracy of Eisenhower's warnings and predictions, and particularly as I thought they applied to "science-based" federal forest management policies. His concerns for the future had been transformed into my concerns for the present.

Eisenhower's speech is rightfully famous for its warning to "beware the military-industrial complex" of centralized government agencies and international corporations in control of our food and chemical production, transportation networks, and armament manufacturing. His principal concerns were that Americans might become subjected to a police state ruled by wealthy and influential elites – one potentially in a constant state of warfare because of profits and political power involved. To avert such an undesirable outcome he called for an ever "alert and knowledgeable citizenry."

This warning, of course, was very familiar to me as with most other public school students in the US during the past 50 years. The part of the speech that was new to me were the following sentences: "The prospect of domination of the nation's scholars by Federal employment, project allocations, and the power of money is ever present and is gravely to be regarded"; and, "we must also be alert to the equal and opposite danger that public policy could itself become the captive of a scientific technological elite."

To my mind this perfectly described my concerns that agenda-based government research funding and the legal profession were being used to develop expensive computer-based models to dictate policy and expand control over the nation's resources. On the surface this could possibly be discounted as a "conspiracy theory," but what alternative theory fits any better? Science has become overtly politicized in the computer age and public policy is being based on the assumptions and manipulations of modelers, "pal reviews," and legal challenges, rather than empirical evidence.

1964 Wilderness Act

I was an incoming freshman at Grant High School in Portland when Lyndon Johnson signed the Wilderness Act in September 1964. The first Wilderness areas totaled 9.1 million acres. The action seemed generally popular at the time, but I can't recall a single student or teacher commenting on it.

This new law recognized "wilderness" as "an area where the earth and its community of life are untrammeled by man, where man himself is a visitor who does not remain." The Act further defined wilderness as "an area of undeveloped Federal land retaining its primeval character and influence without permanent improvements or human habitation, which is protected and managed so as to preserve its natural conditions."

This was an expression of an emerging public desire to "scientifically" manage fragile and uncommon lands and wildlife as protective custodians, rather than as active resource managers.

Over time designated Wilderness areas grew dramatically, including millions of acres of former timberlands, ranches, mines, and fisheries. When historians, anthropologists, and archaeologists began to point out that there was strong scientific evidence that people had been "trammeling" many of these areas for thousands of years – including rock carvings, burials, camas meadows, huckleberry fields, obsidian tools, and more recent traces such as orchards, cabins, fences, and wells – the response was to discount these findings and simply redefine "wilderness."

The author, planting trees in 1981 in Lincoln County, Oregon.
Photo by Bruce Fraser, courtesy Phoenix Communications, Inc.

By the 1980s taxpayer-funded "best available science" somehow began to support "wilderness" concepts of "natural balance," "niche ecology," and "non-declining, even-flow, naturally-functioning ecosystems," in which humans were mostly seen as pathogens and their observed presence only degraded and threatened "the environment." In September 2016, Wikipedia defines Wilderness as "a natural environment on Earth that has not been significantly modified by civilized human activity." Apparently thousands of years of Indian use and occupation of the landscape, followed by a century or more early historical use by European Americans, was accomplished by "uncivilized" individuals.

Today, four federal agencies -- USDA Forest Service, USDA National Park Service, USDI Bureau of Land Management (BLM), and US Fish and Wildlife Service -- administer 759 Wildernesses encompassing 109,754,604 acres.

1966 Endangered Species Preservation Act

In 1966 I was an 18-year-old High School graduate with my first tree-planting job, near Diamond Lake in Douglas County, when Congress passed the 1966 Endangered Species Preservation Act. Whooping cranes, grizzly bears, timber wolves, and bald eagles had been in the news for several years and most people – myself included – thought it was probably a good idea to "conserve, protect, restore, and propagate certain species of native fish and wildlife."

In March 1967 the original "endangered species" list included 14 mammals, 36 birds, 22 fish, three reptiles and three amphibians and had a budget of $15 million per year. The money was to purchase "habitat" for animals on the list.

In 1973, with the strong support of the Nixon administration, Congress almost unanimously passed a completely rewritten Endangered Species Act (ESA). The new law distinguished threatened from endangered species, allowed listing of a species in danger in just part of its range, allowed listing of plants and invertebrates, authorized unlimited funds for species protection, and made it illegal to kill, harm, or otherwise "take" a listed species. In effect, "the law made endangered species protection the highest priority of government."

Today, citing laws and regulations requiring "the best scientific data available," there are more than 2,050 plants and animals listed by the ESA; of which only 28 (1.5%) have ever been determined "recovered" -- including most recently a wolf, a flying squirrel, and four wildflowers. The annual budget is now $251 million per year.

1970 EPA Clean Air Act

In January 1970 I formed a reforestation business with two friends and began performing tree-planting projects for BLM in Coos County; I was a young business owner with a wife and new baby to support. That same month President Richard Nixon signed the National Environmental Policy Act (NEPA) into law, declaring: "the 1970s [will be] a historic period when, by conscious choice, [we] transform our land into what we want it to become" in his State of the Union Address.

It is interesting to consider what Nixon "wanted our land to

Table 1. Large-Scale (10,000+ Acres) Western Oregon Wildfires, 1951-2015

Year	Name	Acres	County	Ownership	Cause
1951	Hubbard Creek	15,600	Douglas	BLM O&C Lands	People
1951	Sardine Creek	21,400	Marion	ODF Santiam State Forest	People
1951	Tillamook IV	32,700	Tillamook	ODF Tillamook State Forest	People
1951	Vincent Creek	28,200	Lane	BLM O&C Lands	People
1966	Oxbow	42,900	Lane	BLM O&C Lands; 1 death	People
1987	Bland Mountain	10,300	Douglas	BLM O&C Lands; 2 deaths	Arson
1987	Douglas Complex I	30,000	Douglas	BLM O&C Lands	Lightning
1987	Silver	96,000	Josephine	USDA Kalmiopsis Wilderness	Lightning
1992	East Evans	10,100	Jackson	BLM O&C Lands	People
1996	Charlton	10,400	Lane	USDA Waldo Lake Wilderness	Lightning
2002	Apple	17,600	Douglas	USDA Umpqua NF	Arson
2002	Biscuit	500,000	Josephine	USDA Kalmiopsis Wilderness	Lightning
2002	Tiller Complex	69,800	Douglas	USDA Umpqua NF	Lightning
2002	Timbered Rock	27,400	Jackson	BLM O&C Lands	Lightning
2003	B&B Complex	90,800	Linn	USDA Jefferson Wilderness	Arson?
2005	Blossom	14,800	Curry	USDA Wild Rogue Wilderness	Lightning
2008	Middle Fork	21,100	Jackson	USDA Sky Lakes Wilderness	Lightning
2008	Rattle	19,800	Douglas	USDA Umpqua NF	Lightning
2009	Boze-Rainbow	16,700	Douglas	USDA Umpqua NF	Lightning
2009	Tumblebug	14,600	Lane	USDA Willamette NF	Lightning
2013	Big Windy Complex	26,700	Josephine	BLM O&C Lands	Lightning
2013	Douglas Complex II	48,700	Douglas	BLM O&C Lands	Lightning
2013	Whiskey	16,200	Douglas	USDA Umpqua NF	Lightning
2015	Stouts Creek	26,500	Douglas	BLM O&C Lands	People

become" in 1970 compared to what we now have, and how people now might "want the land to become" in the future. At that time loggers were still "getting out the cut," and often entire hillsides were "slicked off" during logging operations. There was no such thing as a "riparian buffer strip," and logging and tree planting were routinely conducted to the very edge of rivers and streams. Fishing, hunting, hiking, and camping were generally excellent and new roads were going everywhere the fish and game – and logs and wildfires – were.

The Clean Air Act was also signed into law by Nixon during 1970 and, on December 2nd of the year, he creatXed the Environmental Protection Agency (EPA) to enforce these new regulations. This Agency almost single-handedly created an entirely new industry – one that has continued to grow and thrive to this time: the practice of "environmental law" by hundreds of EPA lawyers responding to the hundreds of legal suits filed by hundreds of lawyers paid by taxpayer-subsidized non-profit organizations.

Computerized "models" of "habitat" and "climate" and "fire return intervals" and the number of government scientists, technicians and lawyers needed to develop, promote, and defend them accelerated rapidly following the creation of EPA. New policies, laws, and regulations sprang – and continue to spring – from their wake. The agency now has an annual budget in excess of $8 billion and employs more than 17,000 people, of which more than half are highly paid engineers, scientists, lawyers, and "policy analysts."

1994 Clinton Northwest Forest Plan

In 1981 my reforestation business was identified as one of the "500 fastest growing businesses in the US" by Inc. Magazine -- the only such business (#332) so identified from Oregon that year. I now had two kids, a small fleet of trucks, more than 30 employees, a new log home of my own design, a few hundred acres of manicured timberland, and represented the reforestation industry at a Congressional hearing about Oregon Wildernesses, chaired by Senator Mark Hatfield. A few years later we were broke, my wife and I separated, and the land and property sold -- just like hundreds of other rural family-owned forestry businesses in the western US at that time.

When I subsequently attended forestry classes at Oregon State University in the late 1980s and 1990s there was a lot of interest in such concepts as "preserving old-growth forests," "maintaining spotted owl habitat," and "riparian enhancement." These concepts were typically rationalized by unfounded theories of "steady state ecosystems" and idealistic descriptions of such circumstances as "non-declining, even-flow, naturally functioning" forests and grasslands. It didn't seem to matter that such conditions had never actually been observed in nature, measured, or documented – only that, for some reason for some people, they were desired.

In 1994 these theoretical "ideals" were integrated into President Clinton's "Northwest Forest Plan" as a "focus on scientifically sound, ecologically credible, and legally responsible strategies and implementation." Since its adoption the plan is claimed to have resulted in a worsening of rural forestry-based economies; a significant increase in the number and severity of large-scale wildfires; and a decrease in several desired native plant and animal species, including deer, elk, spotted owls, oak, and huckleberries.

This table of large-scale western Oregon wildfires tells the story of government science-based policies regarding the management of federal forestlands in the western US.

Even a cursory examination of these numbers is startling: from 1951 through 1986 there were only five large-scale forest fires in western Oregon, an average of one major fire every seven years that cumulatively burned an average of less than 3,900 acres/year; conversely, from 1987 through 2015 there was nearly 2/3 chance of wildfire and an average of 36,700 acres burned per year. That is, for the 29-year period of 1987-2015, large-scale wildfires were nearly five times more likely to occur than during the preceding 36 years -- and averaged nearly 10 times as many acres burned per year. Further, if 1951 is removed from the equation the numbers become even more glaring.

The total is: 24 large-scale wildfires from 1951 through 2015 covering more than 1,200,000 acres. Two 1951 fires took place on Oregon State forestlands -- the remainder all took place on federal lands: six wildfires totaling 154,700 acres on USDA National Forests; 10 wildfires totaling 266,400 acres on BLM O&C Lands; and six fires totaling 733,100 acres on USDA Wilderness lands.

During the entire 1951-2015 period, not a single large-scale wildfire was associated with any of the millions of acres of industrial forestlands and private tree farms in western Oregon. Because climate and native plant and animal species were identical on both types of ownership (government and private), the cause of this stark disparity must be something other than climate or species.

The stark difference in recent fire histories has been identified and discussed in several of my earlier articles as active management vs. passive management. Active management is typified by such activities as road maintenance, vegetation, and wildlife management, and/or recreational developments – all of which took place on federal forestlands from 1951-1966 and continue to take place on private and industrial forests today. Passive management is typified by Wilderness creations, roadless areas, spotted owl habitat, and riparian buffers, as outlined in the preceding pages and characterized by decisions to do very little or nothing -- and even then often restricting uses of motorized tools and transportation.

Dr. Ben Stout, along western shoreline of Round Lake, B&B Complex fire aftermath, May 29, 2004. Photo by author.

2016: Science vs. Modeling

Eisenhower was right. In the years since his farewell address, taxpayer-funded research and related policies and litigation have been accompanied by massive numbers of expensive and restrictive federal regulations requiring huge bureaucracies and thousands of lawyers to enforce. All apparently based on findings and desires of a like-minded corps of government-funded computer-centric modelers and technicians: the "scientific technological elite."

Most of the current policies, laws, and regulations governing our federal, state, tribal, private and municipal lands, waters, and resources are based upon the dictates of these elites. One problem is that many of these policies are based on disproven theories that are inherently racist and strongly biased against past cultures and current populations; a fact that has gone largely unnoticed and unchallenged. Why that is might be hard to explain, but these biases are obviously based more on personal values and political realities than on empirical findings or research. The additional fact that these laws and regulations are claimed and being taught as "science-based" is even more troubling.

The practice of science has been seriously compromised during this process, our rural economies and environments have been significantly damaged as one result. The same argument can be made regarding damages to our native plant and animal populations, the degraded quality of our scenery, air and waters -- as well as the teaching and practice of science itself.

My opinion is that now would be a good time to return to traditional scientific methods to guide our resource management policies; and also to develop a common faith in legitimate experience, earned knowledge, and to better value "intellectual curiosity" over "government contracts," as Eisenhower counseled. It only seems right that we leave our descendants something similar to the wonderful conditions we were given by our own ancestors on the land.

2017 Oregon Forest Fires

Still Predictable and Preventable

Dr. Bob Zybach

In the five or so years I have been writing articles and editorials for this magazine a recurring theme has been that most large-scale and catastrophic forest wildfires in Oregon have been "predictable and preventable." This condition has been particularly true during the past 30 years, and continues to hold true with the current 2017 forest wildfires. And will most likely continue for years to come.

For purposes of this article, and based on my academic research of last century, we can define a "large-scale" forest fire in western Oregon as one involving more than 10,000 acres in size, and a "catastrophic" forest fire in western Oregon as more than 100,000 acres in size. In this sense, "catastrophic" is defined as "a sudden, unexpected event causing widespread change and destruction over the land."

Currently (September 15, 2017) there are nine forest wildfires of 10,000 acres or more in size that have been burning out of control for more than a month in western Oregon (see Table). One result has been the thick, sickly haze that has descended along the most populated portion of the state during the past summer -- along the I-5 corridor from Portland/Vancouver to Eugene, and southward to Roseburg, Grants Pass, Medford, and northern California.

As this is being written, all nine of the listed fires remain out of control and are increasing in size. The Rebel Fire, at 8,700 acres in the Three Sisters Wilderness in Lane County, and the Jones Fire, at 9,600 acres also in Lane County and the Willamette National Forest, are both likely to be added to this list before Fall rains end this year's fire season.

Note: Even though many eastern Oregon wildfires often exceed 10,000 acres and more in size, they are not included in this definition because they typically involve grasslands and shrublands rather than forest trees. The fuels for the 68,100-acre Nena Springs Fire on the Warm Springs Indian Reservation, for example, are described as "grass, dormant brush," while the 52,500-acre Cinder Butte Fire fuels in Harney County are given as "short grass, dormant brush, hardwood slash." In these instances, fire often has a positive, cleansing, and rejuvenating effect on the burned vegetation and there is little, if any, damage to timber or other resources.

Name	County	River	Land Manager	Cause	Date	Size
Chetco Bar	Curry	Chetco	Kalmiopsis W.	Lightning	July 12	188,000
Eagle Creek	Hood River	Columbia	Columbia Gorge	People	September 2	37,600
High Cascades	Douglas	Rogue	Crater Lake NP	Lightning	August 13	60,400
Horse Creek	Lane	McKenzie	Willamette NF	Lightning	August 10	31,700
Horse Prairie	Douglas	Umpqua	BLM		August 26	16,400
Miller	Jackson	Applegate	Klamath NF	Lightning	August 14	34,500
Milli	Deschutes	Deschutes	Three Sisters W.	Lightning	August 11	24,000
Umpqua North	Douglas	Umpqua	Umpqua NF		August 11	41,400
Whitewater	Marion	Santiam	Mt. Jefferson W.	Lightning	August 10	13,700

Table 1. Major Wildfires Currently Burning in Western Oregon, September 15, 2017.

Chetco Bar Fire: 15-Year Jinx?

The Chetco Bar and Eagle Creek fires provide good examples of predictability and preventability – and both for similar reasons. Both fires are continuing to burn and expand as this is being written in mid-September, but the Chetco Bar Fire in particular had been accurately predicted by a long-time acquaintance and self-proclaimed wildfire expert more than 10 years ago, while we were surveying Biscuit Fire damage in the Eight Dollar Mountain and Babyfoot Lake area.

At that time my freind predicted the 2002 Biscuit Fire would reburn in 2017 -- that the fire would take place in late summer and early fall and be driven by an east wind were givens: those have been common characteristics of almost all large-scale forest fires in western Oregon for more than 165 years, since the late 1840s.

This prediction was made somewhat humorously, but led to several good discussions. It was based largely on my own research regarding the "Six-Year Jinx" Tillamook Fires of 1933,1939,1945, and 1951; on the 1987 Silver Complex Fires; and in our common interest in the "predictable and preventable" 2002 Biscuit Fire.

On August 3, while the Chetco Bar Fire was still under 2,000 acres in size and had been monitored for several weeks by paid observers, my friend emailed me: "Chetco Bar Fire, Estimated Containment Date Sunday October 15th, 2017.. Right on schedule. Care to guess the final acreage?"

I replied, jokingly, "17,236." Later I amended this number: "if the wind blows, add a zero. If it's a west wind, cut in half."

Three days later, August 6, the fire had increased to more than 6,100 acres and the email discussion became a little more caustic: "The fire crew is apparently standing around hoping it will blow up so they can get a fat check. 66 people, 0% contained, very little growth. They are doing their east wind dance." By early September the fire had expanded greatly on east winds and exceeded my "10 times" estimate of 172,000+ acres. Fuels were described as:"The area is comprised of mixed conifer and deciduous forest, brush and snags from the previous fires. The fire is burning in areas of fire scar and islands that were previously unburned. The combination of down, dead fuels with a newly cured grass makes the terrain quite hazardous for firefighters."

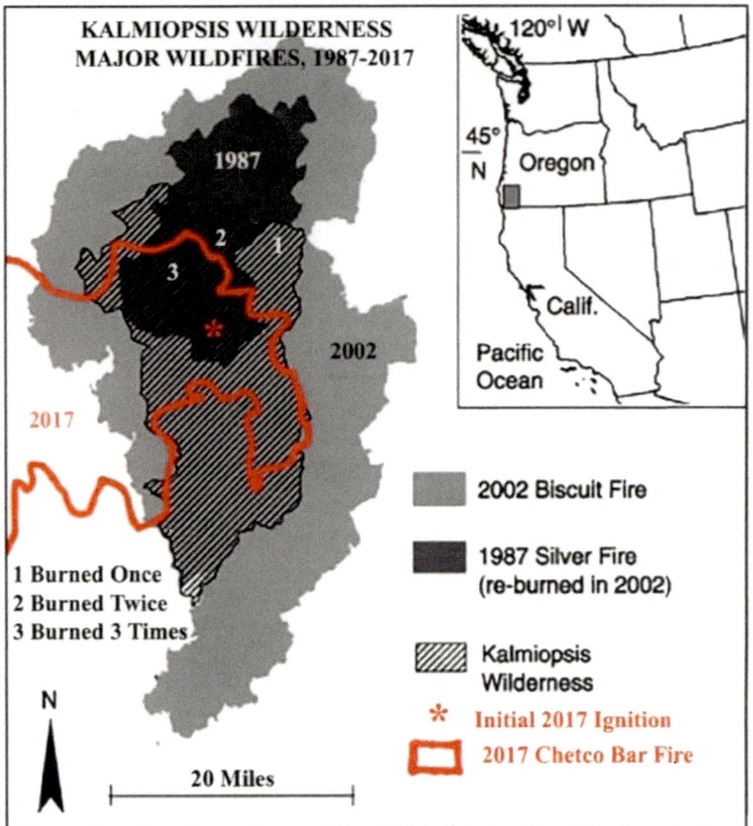

Map of eastern boundary of the 2017 Chetco Bar Fire in relation to 1964 Kalmiopsis Wilderness, 1987 Silver Complex, and 2002 Biscuit Fire. These fires were all predicted due to their fuel loads, location, topography, and changed land management policies.

Eagle Creek Fire: National Scenic Wildfire Smoke

The Eagle Creek Fire was reportedly ignited by kids playing with firecrackers about a mile south of Cascade Locks on September 2. Within a day it had increased in size

In 1989, following the 1987 Silver Complex, I worked with an OSU graduate student team analyzing the fire. The recommendation was to salvage dead trees ASAP in order to gain maximum market value and to reduce the likelihood and severity of future wildfires. Following the 2002 Biscuit Fire, the same recommendations were made and likewise ignored. The current 2017 Chetco Bar Fire was inevitable. And preventable.

Chetco Bar Fire, August 17, 2017, burning in the footprints of the 1964 Kalmiopsis Wilderness, 1987 Silver Complex, and 2002 Biscuit Fire. Photo source: InciWeb.

to over 3,000 acres. As this is being written on September 16, it is 45,600 acres, an increase of 8,000 acres since yesterday. Much of this increase likely has to do with purposeful backfires set to keep the fire from populated areas and to help direct its spread; possibly due to revised estimates as well.

As with the other major forest fires this year, the Eagle Creek Fire started -- and is mostly taking place -- on federal lands: Columbia Gorge National Scenic Area, Mark O. Hatfield Wilderness, and Mt. Hood National Forest. The 1901 Langille USGS Forest Reserve map shows much of this land was classified as "Burned Timber, Not Restocking," "Burned Timber, Restocking," or as contiguous stands of young timber likely growing from past fire events. The photographs, probably taken by Langille at the same time he was constructing his map in 1899 and 1900, illustrate the types of fuels he is representing, and their volatility.

The Eagle Creek Fire is still burning out of control and is continuing to have severe negative effects on the air quality of nearby communities, which include some of the heaviest concentrations of people in Oregon and Washington. Despite this massive amount of air and water pollution, disruption to local commerce and transportation, widespread discomfort and illness, and deaths of thousands of wildlife, local environmental groups are already using this event as an opportunity to raise money and awareness for their organizations!

A significant amount of the fuel contributing to Eagle Creek Fire and pollution can be directly traced to litigation by environmental organizations intended to halt salvage and logging operations in the Mt. Hood National Forest, and to passive management regulations adopted through legislation creating National Scenic Areas, ESA, and Wildernesses.

One of the key players in Mt. Hood-based litigation is BARK. The non-profit organization was created in the late 1990s "to transform Mt. Hood into a place where natural processes prevail." Its principal method for doing so, to date, has been by raising money to pay lawyers to file suits against active management of the mountain's forestlands. Such management actions are not deemed "natural processes" and therefore must be stopped, whereas destructive forest wildfires are considered "natural" and are to be expected, even celebrated. This legal strategy has greatly increased overhead costs to the USFS and American taxpayers, had a serious negative effect on local forest-based businesses and employment, and contributed significantly to the risks, explosiveness, and unhealthy pollution associated with recent wildfires on Mt. Hood, including Eagle Creek Fire.

Yesterday I received an email solicitation from BARK (I usually get several a year) from their "Meeting Facilitator" David Osborn, who is "grateful to BARK for educating [him] about fires." That education has apparently resulted in an understanding that "science shows that weather, not fuel, dictates fire behavior," and that "logging releases more carbon than fires." Thus, we need to write the Governor and our Representative to "not be a science denier!" Instead, we need "to advocate for climate justice" and urge them to "use science about fire, climate change, and logging to inform their policy decisions." I was also invited to participate in their next "Radi-

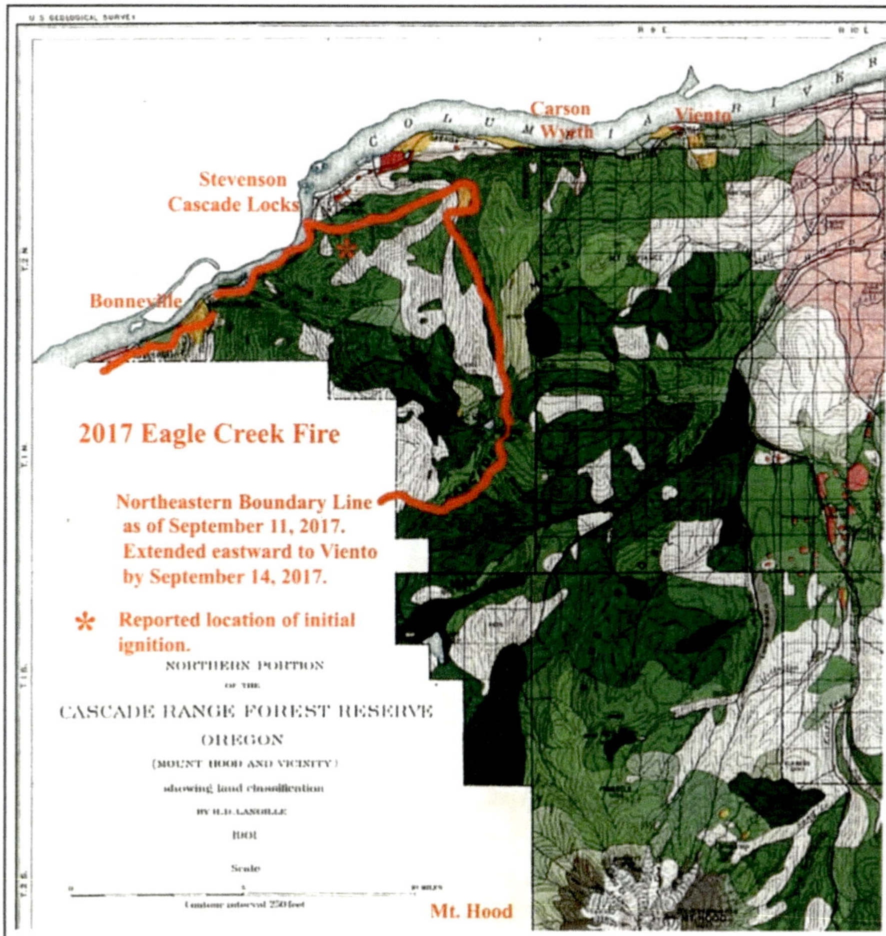

Annotated portion of 1901 USGS map of Mt. Hood forestlands by Harold D. Langille. Note the vertical brown lines in large polygons within the Eagle Creek Fire's boundaries: these represent "Burned Timber, Not Restocking," such as the snags shown in the photo of Ghost-tree Ridge. The remaining polygons likely represent pre-1890 wildfire events as well: the light green areas with vertical brown lines represent "Burned Timber, Restocking," with increasing darkness of green mostly representing increasing volumes of second-growth Douglas Fir that burned and reforested at different times. The scattered darkest green polygons represent stands of "50,000 and Over" [board feet/acre],

Forest Fire on east fork of Hood River. From a photograph taken at Cloud Cap Inn five minutes after the fire started.

Mount Hood from Ghost-tree Ridge. Whitened trunks of trees killed by forest fires.

Photographs from John H. Williams' 1912 book, "The Guardians of the Columbia". Images are from Chapter III "The Forests," written by Harold D. Langille, who also authored the 1901 map of Mt. Hood and may have taken the pictures at that time.

cle Activist Training," titled "Intro to Forest Policy: Forest Fire Edition." Because of my "privilege" and "responsibility."

CO_2 & Global Warming: Smoke Gets in Your Eyes

CO_2 (carbon-dioxide) is a beneficial gas routinely pumped into greenhouses in order to encourage plants to grow. It is odorless and colorless and is considered a "building block of life" on planet Earth. A recent study has shown a nearly 10% increase in the Earth's forestlands during the past 33 years, including 40% greater forest cover in "dry lands" than previously measured. Many scientists are suggesting this apparent increase in forest growth may be due to increased levels of CO_2 in the atmosphere during the same time period.

Other people are concerned that increased CO_2 is largely responsible for catastrophic weather patterns and other consequential and negative effects on the climate. This concern has largely devolved into a political issue during the past few decades: Democrats seem to think that CO_2 is highly dangerous and needs to be regulated by the government for that reason, and Republicans seem to think that such concerns are mostly a paranoid belief not necessarily based on fact. Many people who believe in Global Warming also seem to believe, along with David Osborn, that the evidence is clear and recent wildfires offer proof.

In the past forest fires have killed thousands of people – the 1871 Peshtigo Fire in Wisconsin, for example, is estimated to have killed between 1500 and 2500 people. In the Pacific Northwest, the 1902 Yacolt Fire killed 38 people along the Lewis River in southwest Washington. Today, direct loss of life to people not actively engaged in firefighting is very unusual, due in large part to modern transportation and communication systems.

Instead, the main health problem associated with wildfires today is caused by ash and other particulates in the smoke, which can be very harmful and even life-threatening to young children and older adults with asthma, emphysema, and other breathing-related problems. Even healthy adults can experience painful eye irritations, coughing, sneezing, and other discomforts due to smoke-related effects. Bad tastes, unpleasant odors, and obscured vistas are other problems being reported throughout the western US during the current fire season.

The primary air pollution and public health problem associated with wildfires is not CO_2, it is smoke.

What to do?

The steps needed to reduce the costs, occurrences, and damages caused by Oregon forest fires in Oregon over the past 30 years remain the same:

1) First, we must recognize that virtually all of these fires are taking place – and have been taking place – on federal lands: Wilderness areas, National Forests, Monuments, and O&C Lands. For the most part, they are not taking place on private timberlands and tree farms at all – unless those lands are adjacent to burning federal lands. This clearly shows that these wildfires have nothing to do with "Global Warming" or "Climate Change" and everything to do with management practices. From 1951 to 1987, a 36-year period during which our National Forests were being actively managed and Wildernesses and ESA were just coming into existence, there was only

This photograph went "viral" shortly after being posted anonymously on Facebook and then being posted to Reddit -- where it currently has 116,000 retweets and 229,000 Likes - and then Yahoo News, Golf News Net, and the New York Daily News. On September 7, The Oregonian revealed the amateur photographer as Kristi McCluer, who had not given permission for the Reddit posting and was initially unaware of all the media attention.

The photo is of the Eagle Creek Fire on September 4, 2017, as viewed across the Columbia River, south from the Beacon Rock Golf Course. Because golf courses are actively maintained, seasonally irrigated, and trees are placed individually or in small groves away from one another, they are a very safe environment – except for possible air pollution – from adjacent wildfires. The same is true for irrigated croplands, wetland prairies, many public parks, and other locations with a minimum of flammable structures, flat or rolling terrain, widely spaced trees and groves, and well maintained grasses and shrubs.

one large-scale forest fire in western Oregon. Today there are nine, and counting. The climate has not changed that much in the past 30 years, but management of federal lands has.

2) Environmental organizations have claimed many times that these fires "burn hotter" and are "more damaging" on conifer plantations and young, even-aged stands of trees, as typified by industrial forest management and past USFS and BLM practices. That description fits almost all private forestlands today, yet these areas – despite having the exact same weather patterns as federal lands – remain mostly fire-free during these events. The primary difference between the two is management: private lands are actively managed to protect their investments, while federal lands have been passively managed for most of the past 30+ years.

3) Get to work. There are tens of thousands of family-wage jobs needing to take place on public forestlands in the west, other than just seasonal firefighting: logging, road improvements, tree planting, prescribed burning, campground construction, trail maintenance, etc., and thousands more jobs needed to house, feed, clothe, and transport these workers -- and most of these jobs produce tax revenues, rather than use them. It's a big job but it has major benefits to US society, including rural Oregon, and it needs to be done. Just look at the air.

Help California Rebuild By Managing Our Forests

By Dr. Bob Zybach

Last month I wrote a magazine article regarding this year's catastrophic wildfires in Western Oregon. The article mentioned that in the past, forest fires have killed thousands of people -- the 1871 Peshtigo Fire in Wisconsin, for example, killed an estimated 1,500 to 2,500 people. Today, people not actively engaged in firefighting seldom die in forest fires, due in large part to modern transportation and communication systems.

Within days of the article's publication, wildfires in Northern California killed more than 40 people, burned nearly a quarter of a million acres and destroyed more than 5,700 homes, businesses and other structures. This is the most people ever killed in West Coast wildfire history.

The California wildfires of 2017 are a true catastrophe, whether measured in acres, dollars or human lives.

Western Oregon forest fires also burned more than half a million acres this year. The fires cost millions of dollars to fight, killed millions of wild animals and subjected most of the state to foul-smelling, unsightly, unhealthy, choking smoke in August and September. But no one was killed during these events.

Compared to landslides, hurricanes, floods, earthquakes and volcanic eruptions, the loss of 40 lives might seem relatively minor. For families directly affected, though, it is a true catastrophe. When added to the destroyed homes, jobs, mementos, neighborhoods and communities, the loss must be nearly unbearable.

What can we in Western Oregon do to help rebuild Northern California's homes and communities, and to help reduce the increasing frequency, size, costs and destructiveness of these events?

The best answers likely rest with the Department of Agriculture and the Department of the Interior. Lands managed by those agencies are where the large majority of fires are taking place, and those lands hold most of the dead trees that will likely fuel future fires if they are not removed.

Last month Interior Secretary Ryan Zinke and Agriculture Secretary Sonny Perdue joined forces in directing their agencies to "prevent and combat the spread of catastrophic wildfires through robust fuel reduction and pre-suppression techniques."

These statements were made before the California wildfires took place. Much of their focus was on controlling and suppressing wildfires, but "robust fuel reduction and pre-suppression techniques" imply long-term solutions via active management of our forests. That would signal a major shift from how many federal lands and wildfires have been managed during the past 30-plus years.

From 1951 until 1987, only one forest fire in Western Oregon exceeded 10,000 acres. Since 1987 there have been dozens of such fires, including 10 this year alone. Almost all of these fires have been on federal lands. Very few large fires in the past 70 years have burned on private or state lands -- and those few were mostly affected by adjacent burning federal lands.

The principal difference is that private lands are actively managed for use and protection of their resources, while federal lands have increasingly become passively managed, allowing "natural processes" to take place with little or no human interference.

Active management in forested environments involves timely marketing and salvage of dead and dying trees; selective thinning of some areas; clearcutting, prescribed burning or reforestation in other areas; good road access; maintaining native wildlife populations and providing recreational opportunities. From World War II until the 1980s, most federal forests were actively managed to provide for national defense, post-war housing, wildfire control, public recreation and wildlife habitat. There were only a few large-scale wildfires during those years.

Passive management of federal forestlands largely began with the 1964 Wilderness Act, creation of the Environmental Protection Agency in 1970, and the 1973 Endangered Species Act.

This process accelerated in following decades with government land-use designations, burgeoning ESA and EPA policies, and public litigation resulting in roadless areas, spotted owl habitat, and other large preserves in which active forest management was discouraged or even outlawed.

Recurring large-scale and catastrophic wildfires in these areas has become a predictable result, beginning in 1987 and growing worse since then.

Quickly salvaging trees killed in this year's fires would produce thousands of direct and indirect jobs, greatly improve the economies of rural Oregon and California, rebuild infrastructures harmed by the fires, reduce future wildfire risks and costs, and help provide economical, high-grade construction materials to restore ruined homes, businesses and communities. Safer, more beautiful forests for both people and wildlife would be another important result.

There are humanitarian reasons for shipping finished lumber and other building materials to Northern California as quickly and cheaply as possible. There are sound economic and environmental reasons for doing so as well.

Bob Zybach of Cottage Grove, a forest scientist with a doctorate in the study of catastrophic wildfires, is program manager for the Oregon Websites and Watersheds Project (www.ORWW.org). This was originally published as an editorial in the Eugene Register Guard in October, 2017.

Trouble in Paradise: Why This Fire Should Not Have Taken Place

By Dr. Bob Zybach

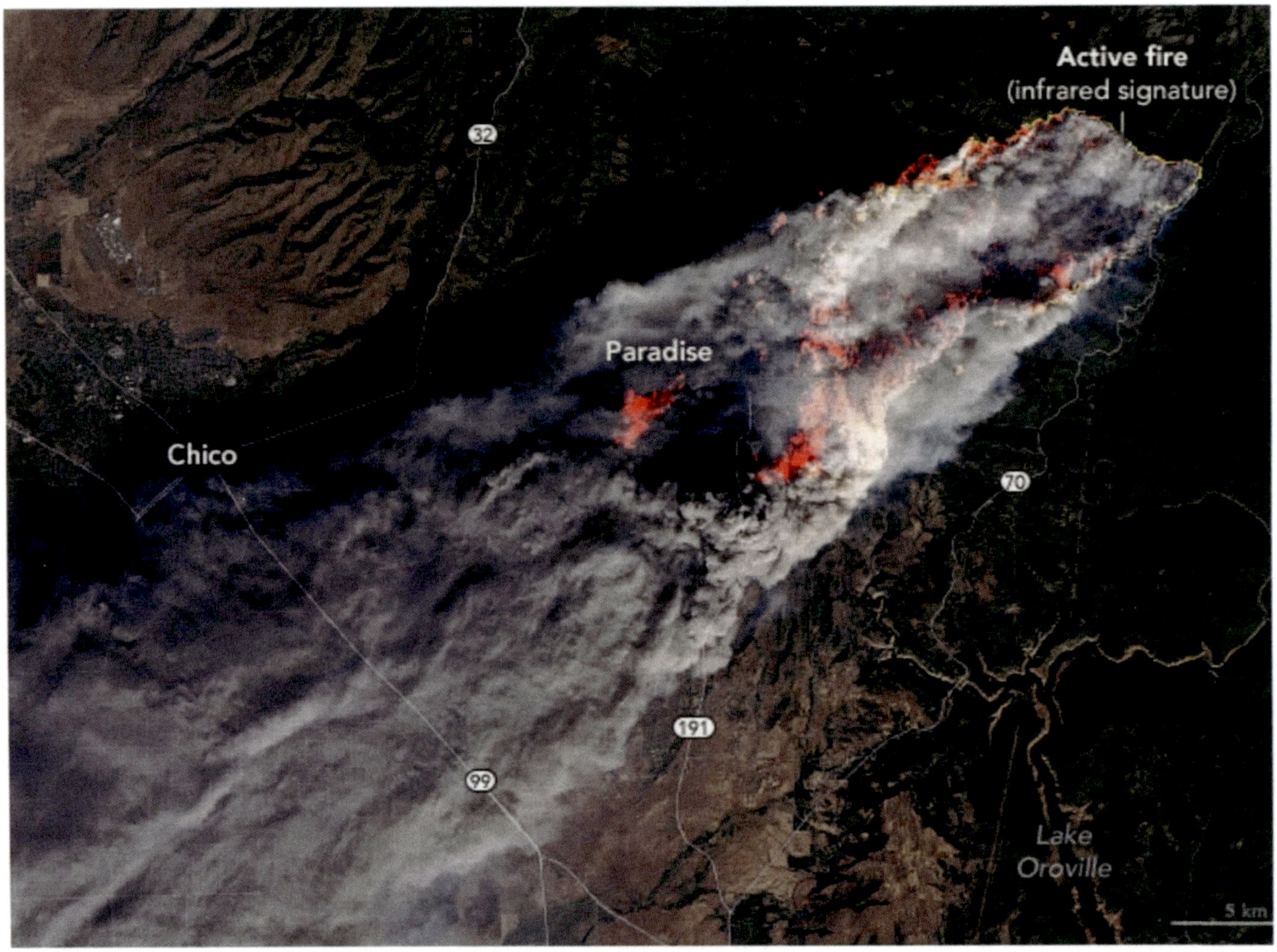

Photo courtesy of NASA.

I have been a co-moderator of, and irregular contributor to, a discussion blog titled "Global Warming Realists" for several years. We have more than a hundred members, including a number of scientists on at least three continents, professional meteorologists, knowledgeable citizens, a few politicians, and at least one member of the media, Lars Larson -- who hosts a popular radio talk show that exists on two levels: his afternoon "Northwest Show" that covers Oregon, Washington, Idaho, and northern California, and his evening "National Show," which also covers most of the remaining 46 states and much of southern Canada.

In early November, as part of a blog discussion regarding the Paradise Fire in northern California, I wrote:

"The area around Paradise had been oak savannah maintained by annual Indian fires for many centuries. Then the Spanish and their cattle moved in and the area was kept free of flash fires by mass seasonal grazing for another century or more. Then the USFS banned cattle grazing, too, and predictable wildfires have followed.

The reason the towns burn so quickly is that they are constructed of kiln-dried lumber — essentially a bone-dry forest dismembered and moved onto a prairie.

Zero to do with "climate change" and everything to do with mismanaged forests and grasslands created by federal regulations on federal lands. This is where junk science and "scientifically managed resources" bump heads."

The next day, Larson quoted this statement on both his Northwest and National programs and we made arrangements for me to be interviewed on the following day.

For many years I have been interviewed by Larson on his radio shows. These interviews usually take place during wildfire season and focus on their cause, effects, and mitigation. We have also discussed other topics, such as spotted owl "habitat" and the management and ownership of the Elliott State Forest, but mostly our discussions have focused on wildfires – and usually as they were taking place, or shortly thereafter.

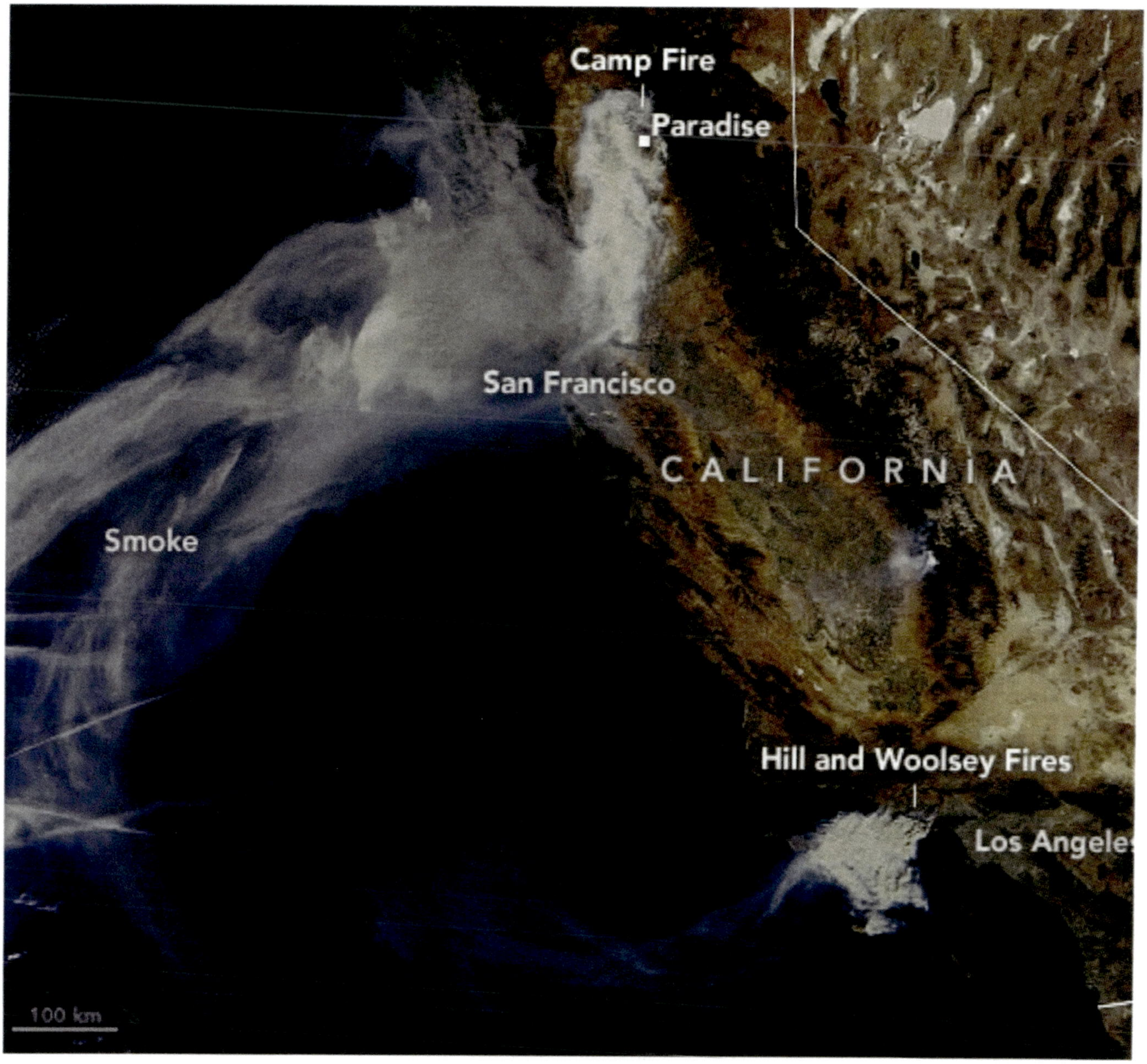

Photo courtesy of NASA

This year's Paradise Fire was the focus for both his regional and national shows on November 14, while the fire was still burning. The following transcript is edited from our conversation to highlight the historical magnitude of this fire, why it occurred, and what can be done to minimize future such occurrences:

Lars Larson: We've been talking a lot over the last week about the fires that began last Thursday, the fires in several places in California. But the most serious, one of the most deadly, in Paradise, California, just up the hill in the Sierra Nevada from Chico, California, the one that's now identified as having killed almost 50 people. Several hundred people are still missing and unaccounted for. And the question is, what caused these fires? Or what could have made this situation better? The president sounded off by blaming the state forestry management, and I think there may be some blame there, but I think most of it is federal forestry management.

But I thought I'd consult with a guy who actually knows the subject well. Dr. Bob Zybach is a forest scientist, president of NW Maps Co., and the author of The Great Fires: Indian Burning and Catastrophic Forest Fire Patterns of the Oregon Coast Range, 1491-1951, his doctoral work that actually covered about 500 years. Dr. Zybach, good to have you back on the program.

Bob Zybach: Thank you, Lars. Good to be here.

Larson: Is there anything that would have made a difference in how the fires turned out for Chico, for Paradise, California?

Zybach: Oh, absolutely. If the fuels had been managed reasonably as had been done for 30 or 40 years by the Forest Service, there would have been a lot fewer fires, including the Paradise Fire. And they would have been a lot less magnitude, a lot safer, a lot less damage, and a lot more tenable.

Larson: Now when you say manage the fuels, you're talking about the wood mass that's grown up in those forests, and of course, the brush and the grass in the areas that are mostly federal lands around Paradise, California -- the same kind of federal lands that we have in much of the northwest.

Zybach: Yeah. Last year we had 10 major fires in Western Oregon. All 10 started on federal land. This year we got the Klondike fire, which is the fourth major fire to come out of the Kalmiopsis Wilderness since 1987. Between 1952 and 1987, there was one fire of this magnitude in western Oregon. It was also on government land in Western Oregon -- but one fire in 30 years when the land is being actively managed, compared to four fires coming out of the Wilderness or 10 fires all occurring on federal land last year are the patterns that are just the same this year.

Larson: Now you might say, I know some people are going to say, "Well, the government can only do so much." But let's go back about 500 years and talk about how the Indians, who didn't have the federal government, who didn't have vehicles, who didn't have any of the rest of this, they managed these forests, and they did it sensibly and actually in a way that kind of mimics nature, did they not?

Zybach: Well, they managed the forests. I don't know about "mimicking nature." Some people think that's an area devoid of humans. But they used broadcast burning and fuel wood gathering, of course, to regularly burn tens of thousands of acres, sometimes to harvest crops, sometimes for hunting, probably sometimes for fun or occasionally even by accident. But that kept the underbrush and the flash fuels under control. It kept fewer stems per acre so the canopies didn't fill in in most areas so densely. So it couldn't get crown fires in many areas. And then a lot of the land such as that in Sacramento Valley or the Willamette Valley was oak savannah. So it was easy to harvest say the tarweed in this area, or acorns, or bracken fern by having large controlled burns for thousands of years.

Larson: Now Dr. Zybach, I'm not suggesting . . . I know some people do, but I'm not suggesting at all that we go out and have fires or deliberately set fires. Or that when a fire does occur, that we just let it burn and say, "That's Mother Nature's way." I know there's some tree huggers who like that.

But I'm suggesting something different that actually could be a money and job and tax revenue maker if we did it. And that is, I think what you've suggested, and that is you go in and you actively do some logging. You take some of the revenue from that and you do some thinning and you do some clearing and maybe you do some controlled burns at the right time of the year when there's enough moisture that you don't get this kind of massive fire. Is that what you're talking about? If not, please correct me.

Zybach: No, that's exactly right. In the '60s and '70s, we didn't have these fires, and we brought in hundreds of millions of dollars off of federal lands through logging. Maybe they were poorly designed logging units by today's standards, but it kept fires down. It brought hundreds of millions of dollars into rural communities and to rural families, and it produced lots of building materials for the housing boom, the baby boomers, and so on. So there was a lot of good, parks, schools in rural communities that came from the income. So this was tax-producing operations rather than firefighting, which is an entire different industry and dependent almost entirely on taxes to support it.

Larson: So Doctor, tell me this. Is there any scientific reason not to go back to that same kind of logging activity, while maybe altering a bit the way it's done so that everybody can be happy and feel warm and fuzzy about it?

Zybach: There's zero scientific reason for not doing so. There are, however, a lot of so-called scientific rationales, whether it's global warming or black-backed woodpecker habitat or marble murrelet habitat. There's excuses and rationales and, I would say, junk science to support these approaches. But biologically, there's no problem at all with active management. And economically, it's wonderful. As far as danger to humans, we've just had a relatively few die or be injured, which is really tragic -- but think

Inciweb (USFS) photo of downed powerlines on November 18, 2018 in Paradise.

of the millions of wildlife; animals that are killed in these events.

Larson: My understanding is the pictures of that coming out of the Paradise area of California are actually horrific. So, Governor Jerry Brown, the lame duck governor of California came right out almost immediately and said, "This is global warming. This is the new normal." Is there any scientific basis for that?

Zybach: No, it's a political basis. The climate's been the same for 500 years. It's weather events in which these fires occur are normal, east winds. It's just the fuel buildup and the management of those fuels, specifically on federal lands and specifically related to federal regulations of the last 50 years, that have predictably resulted in these events.

Larson: So when Governor Brown says, "This is something new," as though Northern California's never had these north and east winds, as though Northern California has never been a very dry place much of the year except the springtime when there's some rain. But most of Northern California is as dry as Eastern Oregon, maybe dryer, even to the point where there's less rain than there is evaporation. So it's very dry. But is that anything new based on the numbers?

Zybach: Nope, totally expected, normal, predictable, average. However you want to look at it, what's new is the fuel buildups: first from the elimination of Indian burning, second from the elimination of massive grazing. In the 1930s, we had what were called the "fern burners." That's one reason we had so many fires in the '30s, because it was a normal practice of people grazing sheep and other livestock in Oregon to fire the prairies after the grazing had been done to rejuvenate them for the following year. So that creates fire breaks and reduces opportunities for a catastrophic fire unless you have a prolonged drought and you have other fuels building up in adjacent lands, which is what also happened in the '30s.

Larson: The other thing I was intrigued by was your description of what a town is when a town is built the way most towns in the West are built, and that is, we build buildings out of stick lumber. We build it out of dimensional lumber that's dried in kilns. So it arrives relatively dried to the building site, and then gets even drier after you put the building together as what little moisture is in it goes away. How did you describe it again?

Zybach: I think a "dismembered forest of bone-dry firewood," was one of the ways I've described it. But it's a forest that's been essentially moved onto prairie lands. Almost all of the major towns in Western Oregon and Northern California are built on Indian prairies. They weren't forest; it wasn't cleared -- it was predominantly prairie land. Then you move all these structures into those places.

One of the things they're blaming the fires on is people moving into the rural areas adjacent to forests and grasslands. It's government, so they give it an acronym, the "WUI" (pronounced "whoo-ee"). Then they create the "wild lands." So those are new designations for flammable areas; poorly managed Forest Service areas, quite often. They're saying all these people are moving into this acronym is the problem. Well, people have always lived there. But now it's 300,000 people or three million people instead of 300. They're bringing in massive amounts of firewood basically; structures built of flammable materials. That's what created the problem in Paradise and in Redding earlier.

Larson: The fires in Paradise actually started quite a distance outside of Paradise. As I understand, almost 10 miles out of town. But it moved very, very quickly once the winds got a hold of it. Is there the potential, perhaps, in situations where you know you've got a town and you got a forest that's very dry that could catch fire, to put in firebreaks in the form of clear cuts that would actually help protect a town by at least providing a defensible barrier between a town and the rest of the forest?

Zybach: Well, I think clear cuts, that's nature's way. Whether it's a volcanic eruption or a wildfire or a Columbus Day storm. It's how nature regularly clears off hundreds of thousands of acres at a time, usually within a matter of less than a week, rather than clear cutting, which might take years. But the real problem is the ladder fuels and the flash fuels in the savannah areas, such as around Paradise. That's an annual problem, or at least biannual.

Larson: The flash fuels are all that dried up buckbrush and grass and everything else that by this time of the year are bone dry.

Zybach: Yep. That's why the fire moved so quickly. Then it gets into town with all that fuel. It's driven by an east wind, and the east wind is coming down off the Sierra Nevada, so it's warmed as it's getting lower in elevation and rapidly drying out the fuels in advance, as well as creating its own weather patterns, which include winds that drive the fire further.

Larson: Dr. Zybach, anything you haven't said already that you would say if, say, Donald Trump put you in charge of the Forest Service, how you'd change things?

Zybach: We'd start making a lot of money, a lot of people would go to work, and there'd be a lot fewer catastrophic fires, a lot less smoke, and a lot more wildlife.

Larson: Unbelievable. Dr. Zybach, it's always a pleasure to have you on the program.

Zybach: Thank you, Lars.

Western Oregon Counties' Forest Fire Histories, 1776-2019

By Bob Zybach, Ph.D

The author is shown discussing historical forest maps at the Round Lake location of the 2003 B&B Complex fire on September 15, 2004. Also pictured are (clockwise from foreground) Kermit Cromack, Jim Peterson, Benjamin Stout (profile), Bill Hagenstein, Zybach, Wayne Giesy, and Nana Lapham

During the past nine years I have written a number of articles about wildfires and their history in Oregon for this magazine. When I began to update that information for this issue on Labor Day weekend, I started by reporting the 2020 fire season was terrible for California, but relatively mild and unremarkable for Oregon and Washington. Then the east winds began increasing in intensity on Labor Day, as I was finishing my first draft.

Today (September 11), news reports are saying that more than 900,000 acres are on fire in Oregon, that ½ a million people are being forced to evacuate their homes, and there have been an uncounted number of human deaths. So far, millions of wildlife have been killed, heavy noxious smoke fills the air, and the towns of Phoenix, Detroit, Blue River, Vida, and Talent have been destroyed.

That quick, the 2020 Labor Day fires have now become the most deadly, dangerous, and destructive in Oregon's history.

As a result, I have changed the focus of this issue's article and will write about Oregon's 2020 fire season for the next issue, after it has ended. I will provide a listing and brief description of western Oregon's previous major fires through 2019 instead, and arrange them in chronological order, by county. This will provide some context for the current events, both historically and geographically.

Western Oregon Counties Bias

There are 18 total counties lying west of the crest of the Oregon Cascades, with Hood River County on the north straddling both east and west slopes. These lands include the locations of some of the largest and most intensive forest fires in world history, many in excess of 100,000 or more acres in size.

The principal reason that these listings are limited to major western Oregon wildfires is because of the types of fuels involved. Many eastern Oregon fires are also tens and hundreds of thousands of acres in size, but typically involve grasslands and shrublands rather than forests, as in western Oregon.

Grassland and shrubland fires have far less fuel and can burn much faster across the landscape than forest fires. In addition, most grasses and shrubs are rejuvenated by these fires and quickly return to pre-fire conditions. Even the forested areas of eastern Oregon – in the Ochoco, Blue, and Strawberry mountains and the eastern slopes of the Cascades – are typically far more open and patchier than their western counterparts, and far less likely to crown.

And it is crown fires that can rapidly kill large swaths of trees for miles on end. This is compared to ground fires, that can actually improve the health of surviving trees by reducing competition for light and moisture and by adding post-fire nutrients to the soil.

The generally deep soils and heavy

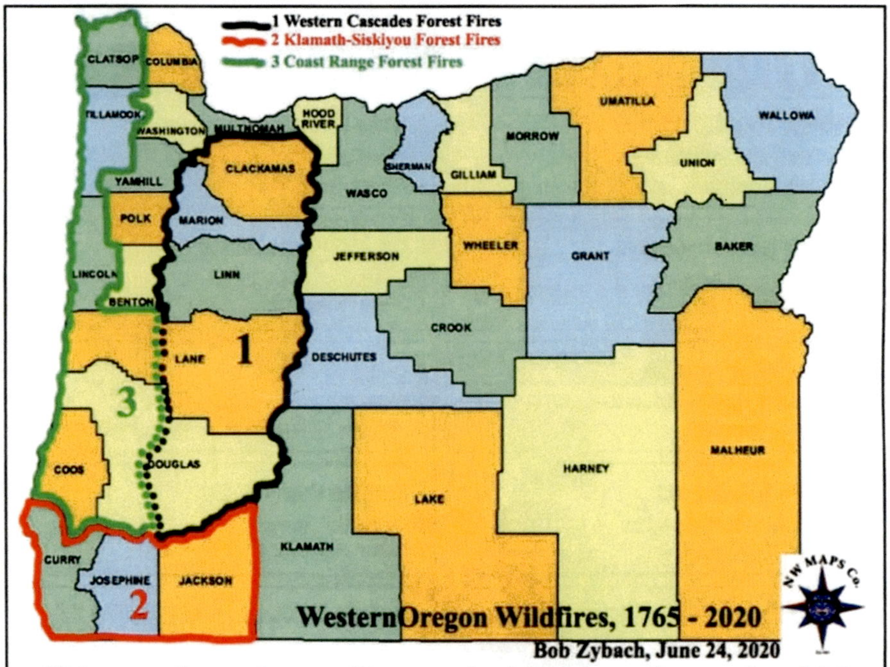

This map shows the specific counties in western Oregon in which major forest fires have occurred during historical time. The three subregions of primary concern are the western Cascades; the western slope of the Coast Range; and the Klamath-Siskiyou Mountains (see table below).

County 1	*Year*	*Wildfire Name*	*Acres*	*County 1*	*Year*	*Wildfire Name*	*Acres*
Clackamas	1901	Barlow Road	85,700	**Josephine**	1931	Humbug-China	10,000
	2011	Dollar Lake	6,300		1987	Silver	96,000
Coos	1776	Millicoma	200,000		2002	Biscuit	500,000
	1868	Coos	100,000		2005	Deer Creek	1,548
	1879	Big Burn	125,000		2010	Oak Flat	7,500
	1929	Fog Belt	10,000		2013	Big Windy	26,700
Curry	2005	Blossom	14,800		2018	Taylor Creek	52,800
	2017	Chetco Bar	191,000	**Lane**	1950	Rasor	5,000
	2018	Klondike	175,300		1951	Vincent Creek	28,200
Douglas	1938	Smith River	28,600		1966	Oxbow	42,900
	1951	Hubbard Creek	15,600		1991	Warner Creek	9,000
	1987	Bland Mountain I	10,300		1996	Charlton	10,400
	1987	Douglas Complex I	30,000		2009	Tumblebug	14,600
	1996	Spring	16,500		2018	Terwilliger	11,600
	2002	Tiller Complex	69,800	**Lincoln**	1849	Yaquina I	450,000
	2002	Apple	17,600		1868	Yaquina II	300,000
	2004	Bland Mountain II	4,700	**Linn**	1856	Canyon Creek	10,000+
	2008	Rattle	19,800		1867	Seven Mile Hill	10,000+
	2009	Boze	10,600		2003	B&B Complex	90,800
	2009	Rainbow Creek	6,100	**Marion**	1865	Silverton	100,000
	2009	Williams Creek	8,400		1886	Silver Cr. Falls	1,000+
	2013	Whiskey	16,200		1893	Scorpion Mt.	9,900
	2013	Douglas Fires II	48,700		1951	Sardine Creek	17,500
	2015	Stouts Creek	26,000		1951	HeeHee	5,000
	2018	South Umpqua	28,700	**Tillamook**	1853	Nestucca	350,000
	2019	Milepost 97	13,100		1918	Cedar Butte	40,000
Jackson	1864	Jacksonville Fires	10,000		1924	Salmonberry I	24,700
	1992	East Evans	10,100		1932	Salmonberry II	43,000
	1994	Hull Mountain	8,000		1933	Tillamook I	311,000
	2002	Timbered Rock	27,400		1939	Tillamook II	209,700
	2008	Middle Fork	21,100		1945	Tillamook III	182,000
	2018	Miles	50,000		1951	Tillamook IV	32,700

This table is a chronological listing and approximate size of major forest fires that have occurred in western Oregon during historical time (see corresponding map above).

seasonal rains of western Oregon are ideal for the growth of Douglas fir, Sitka spruce, true fir, redcedar, and other giant conifers, providing massive amounts of potential fuels per acre, and contiguous flammable crowns in which fires can rapidly spread, and in which tree mortality often approaches 100%.

This is not the case for all western Oregon counties, though. Map 1 divides the region into five subregions: 1) those adjacent to the Columbia River in the north; 2) the eastern Willamette Valley and western Cascades; 3) the western Willamette Valley and eastern slope of the Coast Range; 4) the western slope of the Coast Range; and 5) the Klamath-Siskiyou Mountains in southwest Oregon. Two of these subregions have hardly any history of major forest fires, and the remaining three contain records of some of the greatest wildfires in world history.

Table 1 is a listing of the 64 largest and most deadly pre-2020 forest fires in western Oregon history, arranged chronologically by county. Most of these fires are 10,000 acres or more in size and/or included human fatalities. The acreage figures for several of these fires are best estimates, placeholders for unknown sizes, or reasonably selected from a number of varying sources.

[NOTE: County historians, please check this data. You know your county's history better than I do – am I missing any major wildfires? Do you have better acreage figures or maps? All help appreciated!]

Columbia River Counties

The western Oregon counties adjacent to the Columbia River are Clatsop, Columbia, and Multnomah. There is no record of a large-scale forest fire starting in any of these counties. A lone exception might be the 50,000-acre 2017 Eagle Creek Fire, that started in Hood River County and moved westward into Multnomah County, on the border between eastern and western Oregon. See my article in the Fall 2017 issue of this magazine, "2017 Oregon Forest Fires: Still Predictable and Preventable," for a more detailed description of this fire.

Clatsop County has also had portions of catastrophic wildfires extend into its boundaries, but the nearby 6-Year Jinx of 1933, 1936, 1939, 1945 and 1951 Tillamook Fires mostly started and took place in Tillamook County.

The principal reasons that large-scale forest fires rarely occur on the western Oregon side of the Columbia River are the relatively small sizes of the counties; the great amount of humidity adjacent to the Pacific Ocean and tidal Columbia River; the topography; and the history of relatively dense human populations.

The forested lands adjacent to the Columbia River are typically steep and narrow, draining the northern slopes of Mt. Hood, the Tualatin Mountains, and the Coast Range. These constituted some of the earliest old-growth logging in Oregon, and rare spot fires rarely extended beyond the nearby ridgelines.

Archaeological evidence and early historical records document the Columbia River has been a major trade route and fishery for millennia, and necessarily heavily occupied for those reasons alone. Islands and floodplains became campgrounds, villages, towns, and cities, requiring enormous amounts of building materials, boats, canoes, and firewood on a daily basis. The processes of constantly gathering fuels and maintenance of open areas for human occupation further reduced the likelihood of wildfire of any size.

Western Cascades Counties

Kalapuyan man and eastern Coast Range foothills drawn in 1841 by Alfred Agate, a member of the Wilkes Expedition, near present-day Monroe, in Benton County. Regular landscape-scale fires set by Kalapuyan families and hundreds of generations of their ancestors on the land resulted in open grasslands and oak savannah - rather than forests - throughout most of the Willamette Valley and eastern Coast Range.

The western Cascade Mountains have been the location of a number of major forest fires from their earliest history, beginning with fires documented by the builders of the Barlow Trail along the southern flank of Mt. Hood in 1845 and 1846.

The mythical "Silverton Fire" around the time of the Civil War was said to have been a million acres in size, but modern research has failed to show evidence of even 100,000 acres having burnt near that location in the 1860s, and maybe as a result of two or three major fire events, rather than just one.

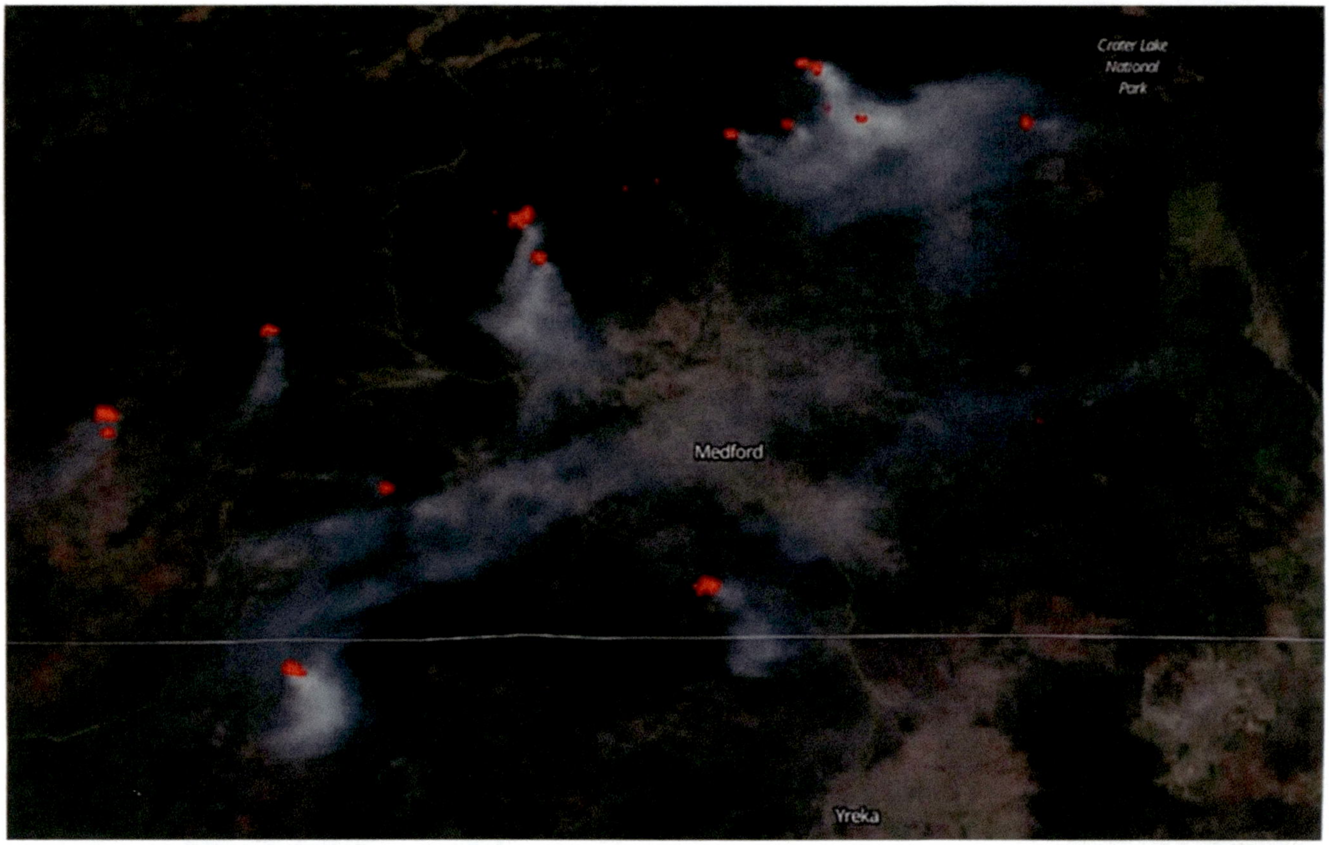

NASA satellite view of southwestern Oregon wildfires ignited by July 15, 2018 lightning storm. This image was created on July 18 and captioned by Lynn Jenner the following day.

Instead, until Labor Day weekend, Clackamas and Marion counties have not had major forest fires since 1951; Linn County suffered the predicted B&B Complex fires in 2003 (see Figure 1); and Lane County has had a handful of large, but not overwhelming, events. Now these four counties are experiencing the Riverside, Santiam Canyon, and McKenzie River fires all at the same time, covering hundreds of thousands of acres, tens of thousands of evacuations, and significant losses of both human and wildlife lives. The communities of Blue River, Vida, and Detroit have been destroyed and others, such as Gates, Mill City, and Lyons, have been significantly damaged.

Lane and Douglas counties are the only two in Oregon that include both western Cascades and western Coast Range landscapes. Until 1987, the majority of major forest fires in these two counties had taken place on the Smith and lower Umpqua rivers on the Coast Range. From 1987 until now, wildfires from 10,000 to 70,000 acres in size have become almost annual events within their Cascades boundaries.

It is worth noting that almost all of the fires listed in Table 1 from 1987 until 2019 have taken place on federal properties: designated Wildernesses, US Forest Service, Bureau of Land Management lands. This is not a "climate change" issue as these events are not taking place on private and state lands that continue to be actively managed have the exact same weather patterns.

East Slope Coast Range Counties

Washington, Yamhill, Polk, and Benton counties include the western Willamette Valley and the eastern slopes of the Coast Range. There is no history of forest fires in these counties that approach even a few thousand acres in size. That is almost entirely because forest trees have only recently begun to invade the prairies and oak savannahs that white immigrants found when they began arriving in the 1830s and 1840s.

Figure 2 is a Kalapuyan man and the local eastern Coast Range foothills drawn near present-day Monroe, in Benton County. Notice the bare hills, few conifers, and missing lower limbs – whether removed by regular burning or gathered for other reasons. The primary reason that these counties lack a history of forest fires is because they lack a history of forests.

Subsequent settlement by US immigrants used these lands for grazing thousands of cattle, sheep, and horses before converting them to towns, cities, fenced farmlands, and housing developments.

Lane and Douglas counties also contain a certain amount of Willamette Valley and Umpqua Valley grasslands and east slope Coast Range hillsides, but they are separated by the Calapooia Mountains, which are generally so jumbled topographically that large-scale fires rarely occur there, either.

West Slope Coast Range Counties

Unlike the eastern slope of the Coast Range, with its history of Indian burning, grazing, agriculture, and large human settlements, the western slope is the site of some of the largest and most famous catastrophic-scale forest fires in history.

In 1912 noted US Geographer and Theosophist, Fred Plummer, authored "Bulletin 117 of the Forest Service," in which he identified the 15 "greatest forest fires in the United States since the year 1800." Three of those fires – 20% of the total -- were in Tillamook, Lincoln, and Coos counties. These fires were followed by the 1933-1951 "6-Year Jinx" Tillamook Fires, which became the best documented and most publicized of all the "Great Fires," due to the existence of film, photography, and international press at that time.

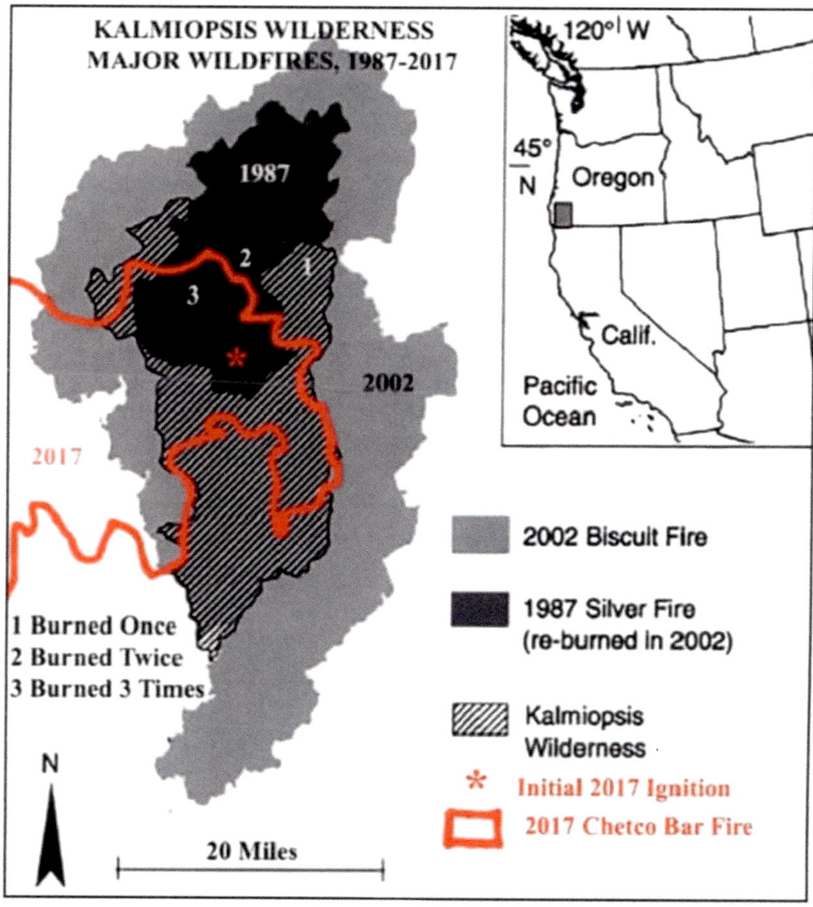

This map of the 1964 Kalmiopsis Wilderness Area in southwest Oregon shows that a number of large-scale forest fires are created by burning the dead, dry trees, shrubs, and other fuels from earlier fires. The map doesn't show the 2018 Klondike Fire oundaries, which also occurred in the same area.

A little-known fact regarding these widely recognized fires is that trees burned during the original fires became the basis for even greater future fires. The 1849 Yaquina Fire and the 1853 Nestucca Fire both expanded their range and severity during the 1868 Fire Year; and the 1868 Coos Fire became larger and more destructive as the 1879 Big Burn. The series of Tillamook Fires were not an aberration, they were the norm (see Map 2).

From 1952 until 1987 the only major forest fire to occur in western Oregon was the 43,000 acre Oxbow Fire in western Lane County.

Lightning is very rare in the Coast Range, and when it does appear it is almost always accompanied by heavy rains. A relatively minor lightning-caused wildfire was noted in 1928, but all of the major forest fires on record are known or believed to have been started by people.

Klamath-Siskiyou Counties

Curry, Josephine, and Jackson counties in southwest Oregon contain the Klamath and Siskiyou mountain ranges, which they share with northwest California. Unlike the Coast

Range, the vast majority of wildfires in this subregion are caused by lightning. Another difference is that much of these counties are covered with shrubs and grasses as fuels, rather than solid stands of trees.

Figure 3 shows the pattern of lightning strikes that resulted in a number of fires on July 15, 2018. Table 1 gives a listing of the largest of these fires.

A similarity for all western Oregon subregions is that arsonists also set fires. The deadly 2020 Almeda Fire, for example, was started in this way, and is responsible for the destruction of the towns of Phoenix and Talent, in addition to the loss of human lives.

Map 2 illustrates the predictable "repeat fires" that have involved the Kalmiopsis Wilderness Area. This is the same pattern of trees and shrubs killed by a wildfire producing greater and more flammable fuels for future fires. This pattern has existed throughout western Oregon for all historical time.

Conclusions

The general information provided by the timing, extent, and location of these major wildfires should be of interest to western Oregon resource managers and US taxpayers -- and to their elected representatives. Here are some basic conclusions that can be drawn from these events:

1) Each county has its own unique history of large-scale wildfires, with significant differences between them: e.g., Benton County has never experienced a large-scale forest fire; Tillamook County has had numerous such fires from 1853 until 1951, and little or nothing to the present time; while Douglas County had few major fires until 1987, and have seemingly had them on an almost annual basis ever since.

2) There were hardly any major wildfires in western Oregon between 1952 and 1987; a 35-year period in which these forests were the most actively and intensively managed in their history.

3) Almost all major wildfires during the subsequent 33 years, from1987 to 2019, have occurred on federal lands – rather than private, county, or state -- and were mostly ignited by lightning or arsonists.

Recommendations

The dead and dying trees of the 2020 Oregon Labor Day Fires should be sold and harvested as quickly as possible. Current log prices are high, family-wage jobs are sorely needed, this would create a much safer, more aesthetically pleasing environment for people and wildlife in coming years and decades.

Reforestation planning on salvaged areas should consider historical prairies, meadows, trails, campgrounds, and riparian vegetation patterns in their implementation, as well as reducing risk of future catastrophic wildfires by uses of prescribed burning and other methods to maintain these environments.

Western Oregon's 2020 Labor Day Fires

By Bob Zybach, Ph.D

Oregon's 2020 Labor Day wildfires are the worst in the state's history, covering more than a million acres, destroying more than 4,100 homes, causing the emergency evacuations of more than 40,000 people, and killing millions of wild animals.

At least nine people died in these fires and most of western Oregon was blanketed with unhealthy, unsightly, acrid smoke for nearly two weeks. For much of that time this air pollution was considered the worst for the entire planet and the long-term health effects of these conditions are still being evaluated.

How did this happen, and how can we keep this type of event from happening again? Or is that even possible?

With a single exception, the 1966 Oxbow Fire, there were no major wildfires in western Oregon for 35 years, from 1952 until 1987. Since then, during the most recent 34 years, there have been nearly 30 such events, with three large-scale (10,000-plus acres) and five catastrophic-scale (100,000-plus acres) wildfires taking place this year alone. Two other fires, Echo Mountain and Almeda Drive, totaled more than 8,000 acres, destroyed 2,700 homes, and killed at least three people.

All 10 of these deadly fires were tied to atypical east winds that began on Labor Day, September 7, and continued intermittently through September 10. What changed between 1952 and

The upper photo was taken with an Osborne camera from the Kinney Ridge Lookout on June 4, 1937, a few miles west of Detroit, looking downstream along the North Fork Santiam River, toward the current location of Detroit Dam and Reservoir. The two lower photos were taken by Mattija Weeks, Asplundh Tree Services, while clearing right-of-way for the PGE powerlines near the mouth of Breitenbush River, a short distance upstream from Detroit. The picture on the left was taken on July 9, 2020, and the picture on the right was taken on October 12, 2020.

1987; and from 1987 to 2020? From one major fire in 35 years to nearly one major fire – or more – a year?

A wildfire needs four elements in order to occur: weather, topography, fuel, and a source of ignition. Topography is mostly a constant; weather has been largely predictable on a seasonal basis for thousands of years; ignitions take place wherever people or lightning – or both – are present; and fuels are the factor that are most often moderated by human actions and decisions.

Sources of ignition.

In western Oregon most wildfires are started by people, followed by lightning, and then volcanoes. Lightning and volcanoes are not a factor on the Coast Range, leaving people as almost exclusive causes of fire in that region; lightning-caused wildfires are also common, along with human-caused fires, in the Klamath-Siskiyous and Cascades; and volcanic fires only take place in the Cascades – and then very infrequently. People and lightning are the most common causes of wildfire there, too.

What is of great interest is the timing, severity and widespread distribution of the Labor Day Fires: eight of the 10 named fires shown on the map were reported during the evening of September 7 through the early morning of September 8. The Holiday Farm Fire was first reported at 7:30 PM on Labor Day, followed by Echo Mountain at 11:45 PM, Riverside at midnight, South Obenchain at 1:59 AM, Slater at 6:43 AM, Archie Creek at 7:37 AM, Thielsen at 8:30 AM, and then, at 11:00 AM, an arsonist started the Almeda Drive Fire.

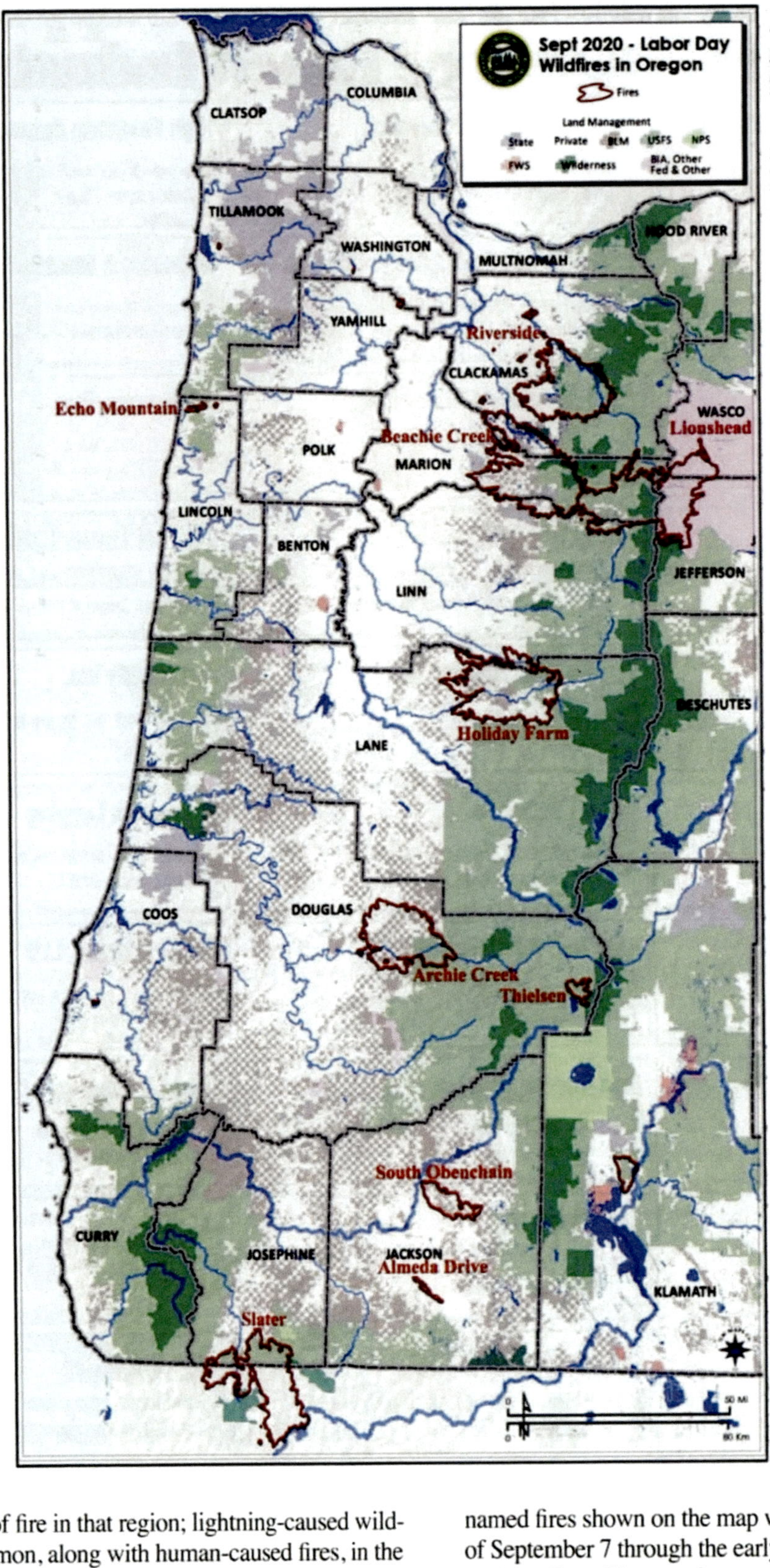

Almeda Drive. Jackson County. 5,700 acres, 2,400 homes, 3 deaths.

Archie Creek. Douglas County. 131,500 acres, 138 homes, 1 death.

Beachie Creek/ Lionshead [Santiam]. Marion County. 300,700 acres, 629 homes, 5 deaths.

Echo Mountain. Lincoln County. 2,600 acres, 293 homes.

Holiday Farm [McKenzie River]. Lane County. 173,000 acres, 517 homes, 1 death.

Riverside. Clackamas County. 138,100 acres, 62 homes.

Slater. Josephine County. 34,000 acres (Oregon); 700 homes, 2 deaths (California).

South Obenchain. Jackson County. 32,700 acres, 33 homes.

Thielsen. Douglas County. 10,000 acres.

GIS Mapping by Teresa Zena Alcock, Oregon Department of Forestry, November 18, 2020.

Statistical data courtesy of Mason, Bruce & Girard, December 8, 2020.

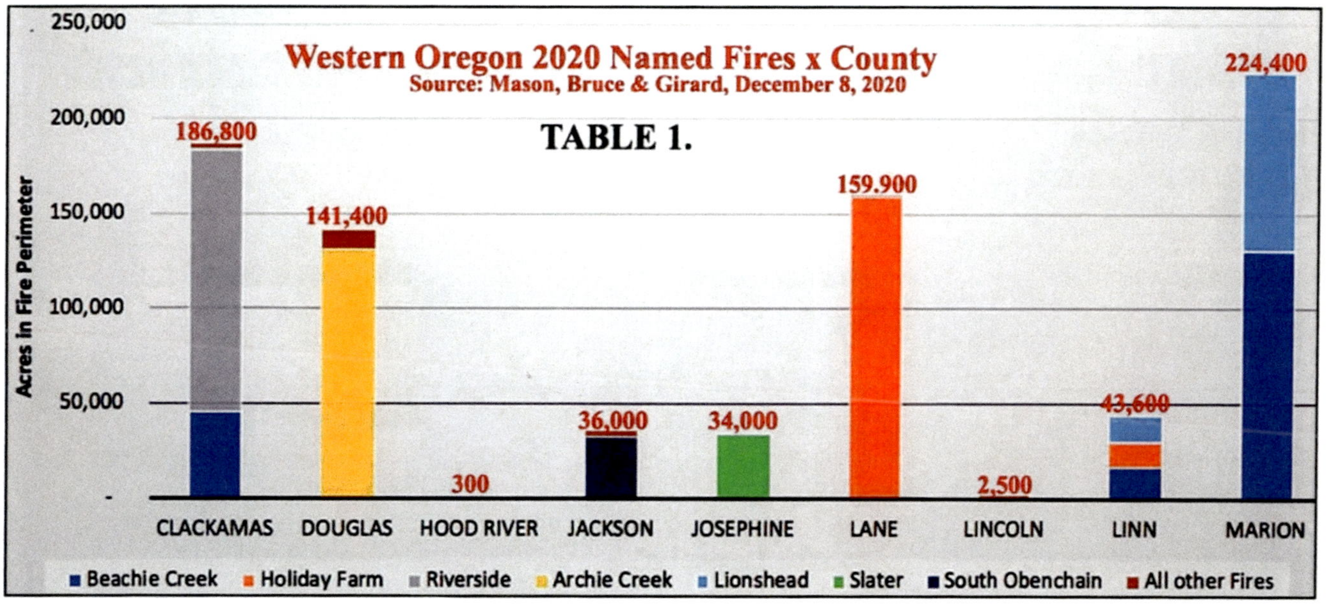

Curiously, investigators listed all but the Almeda Drive event as being ignited by "unknown" causes; or remain "under investigation" as this is being written in mid-December. Not surprisingly, much speculation is that the fires were mostly started by deep-pocket utility companies, or – given the suspicious timing of the fires – by political arsonists sympathetic to the Portland rioters.

Two of the fires that also blew up during those hours actually started on August 16: the Lionshead Fire by lightning on the Warm Springs Indian Reservation in the eastern Cascades, and the Beachie Creek Fire by an unknown cause in the Opal Creek Wilderness in the western Cascades. The strong September east winds quickly caused the Lionshead Fire to burn completely through the northern part of the Mt. Jefferson Wilderness and connect with the Beachie Creek Fire to destroy or greatly damage the Marion County towns of Idanha, Detroit, Gates, and Mill City.

One generally unrecognized factor in the great number and widespread causes of the Labor Day Fires may have been prolonged stay-at-home orders and related unemployment due to the coronavirus pandemic. These fires all started or blew up on the third day of a three-day weekend in which people were widely reported as recreating outdoors – many for the first time in months – on a beautiful day near the end of summer.

Topography.

Major wildfires rarely spread widely over flat lands due to a variety of reasons. One major reason is that these lands are most often occupied by people and livestock and available fuels are kept to a minimum by firewood gathering, plowing, irrigation, mowing, paving, and grazing.

Homes and other flammable structures are a separate consideration, and particularly if constructed from wood and built or placed in close proximity to one another. The Almeda Drive Fire that largely destroyed the towns of Phoenix and Talent this year would be one example, and the mobile home parks of Rainbow, Blue River, and Echo Mountain would be others.

Fires moving uphill are often the hottest and fastest moving, traveling on ambient air currents and heating and drying the fuels in advance of the flames – which are typically longer and hotter than those of fires moving downhill or across flat surfaces. However, the Labor Day Fires were driven downhill from the Cascades crest, driven by the unusual strength of the east winds.

Weather.

Catastrophic-scale wildfires in western Oregon all travel on an east wind. East winds are continental and dry out vegetation as they go; west winds come in from the Pacific Ocean and typically carry great amounts of moisture in the forms of rain, snow, and humidity. As a result, east winds tend to spread fires and west winds tend to suppress or extinguish them.

East winds taking place in the summer months of July, August, and September are the ones most associated with catastrophic wildfires. The winds of September 7, 8, and 9 were unusual in that persistent gusts often reached 50 or 60 miles-per-hour, causing the widespread fires to literally explode.

Prior to 2020, the best known example of this type of concordance of fire and a strong summer east wind was the Tillamook Fire of 1933 – which blew up at midnight August 24 with "gale force" east winds and burned more than 250,000 in the next two days, before being stopped by west winds, fog, and rain. A very big difference is that the Tillamook Fire was a single event, whereas the Labor Day Fires were taking place from the California border to the Columbia River.

According to meteorologist Chuck Weise, weather events similar to August, 24-25, 1933 and September 7-9, 2020 also occurred on September 15-17, 1965, September 14-17 and 25-27, 1970, and September 14-17, 1971 – yet none of these latter events are associated with large wildfires of any size or consequence.

If these events are "caused" by weather or somehow associated with "climate change," as many have claimed, why were there no catastrophic wildfires in 1965, 1970, or 1971? Weather, topography, and widespread use of fire by people on a strong east wind were all the same, or very similar. The principal dif-

Top Left: Ground Fuels. Top Right: Ladder Fuels. Bottom: Crown Fuels.
Photographs by McKenzie Peters, NW Maps Co., Mt. Thielsen Trail, August 23, 2020.

ference would seem to be fuels.

Fuels.

The catastrophic wildfires of the past 30 years have been generally and specifically predicted by me and others based almost entirely on fuel patterns, and particularly on those patterns that have been developing since the 1960s on federal lands.

Wildfire fuels and related events are closely related to ownership patterns. With few exceptions, almost all major wildfires

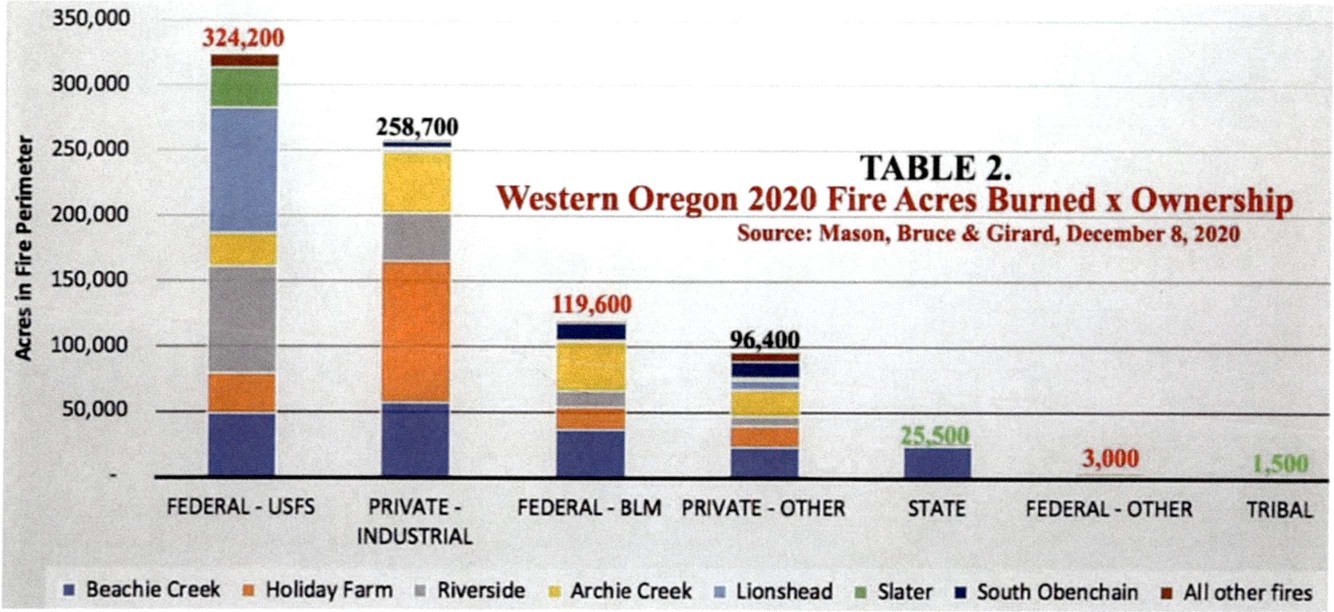

in western Oregon the past 70 years have taken place on federal lands, including designated Wilderness areas, National Forests, and BLM timberlands.

The Labor Day Fires in general, and the Holiday Farm Fire in particular, are something of an anomaly in this regard by being largely fueled by actively managed timberlands and small towns. Still, western Oregon lands that burned this summer included 445,000 acres of federal forestlands as compared to 335,000 acres of privately owned residences, family tree farms, and industrial forestlands.

The apparent reasons for this disparity, and for the predictability of these events, can be found in the series of federal regulations and environmental law determinations since the 1960s. By political decree and legal decisions, the post-WWII actively managed federal forests have been incrementally sidelined as passively managed wilderness, roadless areas, streamside buffers, and "critical habitat."

These actions and decisions have resulted in massive forest fuel build-ups in these areas, and catastrophic wildfires have predictably followed. This has nothing to do with unusual weather conditions or a changing climate, and everything to do with unprecedented and unstable accumulations of flammable vegetation over millions of contiguous acres.

Three of the photos accompanying this article were taken along the southern perimeter of the Thielsen Fire about a month before it burned. They clearly illustrate the problems currently taking place on federal lands – the increasing accumulations of dangerous ground, ladder, and canopy fuels in our public forests:

Ground fires are much like grass or shrub fires, in that they often rejuvenate and protect the remaining stand of trees if regularly burned. This can be a beneficial process, and particularly when regularly completed under controlled circumstances. However, when ground fuels are allowed to build up over time they can burn so hot as to girdle and kill, or severely damage, adjacent trees

Ladder fuels are the lower limbs, intermediate-sized trees and shrubs, hanging mosses, and dead wood that carry the flames of a ground fire into the crown of a tree.

Canopy, or crown fires are the main cause of tree mortality during wildfires. In Douglas fir, a crown fire can readily kill 90% or more of a contiguous stand of trees. The resulting snags are of most concern to forest managers, or should be. They are ugly and dangerous, likely fuel for future fires if left to remain,

These photos of Mount Thielsen and Thielsen Creek subbasin were taken from the Cinnamon Butte Fire Lookout. The upper photo was made with an Osborne camera on August 9, 1937. The lower left photo was taken on August 3, 2020 by McKenzie Peters, NW Maps Co., and the lower right video-clip, also by Peters, was taken from the same general location on October 27, 2020. It is interesting to compare the forest patterns in 1937 with the 2020 Thielsen burn patterns; in particular, the distant linear burn pattern and the foreground trees and cuttings.

yet can often be profitably marketed if sold and salvaged within a short time of being killed.

What next?

These deadly and destructive events were predicted and could have largely been avoided if active management policies had remained in effect on our federal lands.

Fire-killed trees rapidly lose value through deterioration, and also due to glutting the market when massive numbers of logs are harvested at about the same time. There is a real need for rural jobs, reconstruction and repair funding, and structural building materials following the destruction and devastation of the Phoenix, Talent, Rainbow, Blue River, Vida, Gates, Mill City, and Otis communities.

Dead trees pose a real and proven danger to the communities just named, but also to the larger cities of Portland, Salem, Eugene, Roseburg, Medford, and Lincoln City to the immediate west of those towns. The primary danger is wildfire and smoke driven by east winds, as historically shown by the multiple fires and reburns in Tillamook County from 1933 to 1951 and in the Kalmiopsis Wilderness from 1987 to 2018.

This urgency is not lost on private landowners – they have already started salvage logging operations. The problem is on state and federal lands, in which minor logging sales have been held up for years by legal actions initiated by environmental organizations and their legal staffs.

Following harvest of burned lands, the next steps include site preparation, reforestation, and stand maintenance of the rejuvenated forests and woodlands. Improved road and trail access, timely thinnings, and reintroduction of prescribed burning should be important aspects of these strategies.

The time has long passed in which state and federal rules and regulations regarding management of our public forests should be revisited and updated. These fires were largely fueled by Wilderness areas, neglected timberlands, artificial "critical habitat" designations, and overgrown "riparian zones." It will be interesting to see what, if anything, happens as a result of this unnecessary destruction.

The Snags of the North Umpqua:
Part 1. Future Risks

By Bob Zybach, Ph.D

A snag is a dead, standing tree. When it falls to the ground it becomes a log. Snags can be created from living trees struck by lightning, killed by bugs or disease, purposefully girdled or poisoned by people, or burned in a fire. Once a snag is created, the question becomes whether to leave it standing, or turn it into a log.

In western Oregon the most common tree is a Douglas fir. It typically grows in vast stands of even-aged trees numbering dozens or hundreds per acre. Individual trees can grow and survive for centuries, even when the surrounding stands have been decimated by windstorms, floods, landslides, spot fires or other means. However, when a "crown fire" goes through a stand of Douglas fir, nearly all (or all) of the trees are killed and instantly become snags.

When they are living, Douglas fir are among the largest trees in the world and typically include massive amounts of water in their trunks, limbs, and needles. When they die, the water evaporates and the great amount of pitch in these trees -- without water -- becomes far more flammable.

Photograph by Jeff McEnroe, Bureau of Land Management.

One result is that Douglas fir, when dried, makes great firewood: easy to start, burns hot and bright, but throws sparks. Another result is that large Douglas fir snags quickly become great, tall tubes of pitchy air-dried firewood, often capable of casting firebrands, cinders, sparks and smoldering duff a mile or more when being burned during a wildfire.

2020 Archie Creek Fire

The September 8, 2020 Labor Day Fires in western Oregon burned nearly a million acres of land in a three-day period, killed 11 people, destroyed more than 4,000 homes, polluted the air with toxic smoke for nearly two weeks and killed millions of native wildlife.

Of this amount, the Archie Creek Fire, along the North Umpqua River in Douglas County, was responsible for one human death, more than 150 people losing their homes, and over 131,000 acres, mostly forested, being burned. Due to its large size and rapid spread, mortality of native plants and animals approached 100% within much of the fire's perimeter (see Figures 1 and 2).

The Labor Day Fires exploded on strong east winds in western Oregon and took place from the California border to the Columbia River. Fuels included millions of snags from earlier fires, crowded mobile homes and trailer parks, Endangered Species Act (ESA) "critical habitat" designations, regulated streamside "buffers," roadless areas, industrial plantations, even-aged stands of conifer old-growth (trees older than 200 years) and 2nd-growth (trees less than 200 years). The "even-aged" conifers, including snags and plantations, were mostly Douglas fir.

These fires, including the Archie Creek Fire, were typical of catastrophic-scale western US wildfires of the past 35 years in that they predictably began in, and grew from, federal lands (see Map 1 and Table 1), including: Wildernesses, National Forests, and Bureau of Land Management (BLM) range and timberlands.

They were atypical in that they also spread rapidly through thousands of acres of industrial timberlands, family farms, recreational developments and rural communities (see Map 2).

Figure 3 is the view east and southeast of the historic Mount Scott Lookout, taken in August 1933. This lookout had been in operation since 1914 but was destroyed in the Archie Creek Fire. Figure 4 shows the same directional view from the same location, as taken by a "detection camera" in August 2021, magnified 1.5 times.

It is interesting to note that the second-growth trees in the 1933 photo that block the view of Chimney Rock, may have formed the old-growth snags in the 2021 photo, due to the poor growing conditions for Douglas fir on the east slope of Mount Scott. Of additional interest is the fact that these snags likely pose little future risk due to their remote setting, small size and wide spacing.

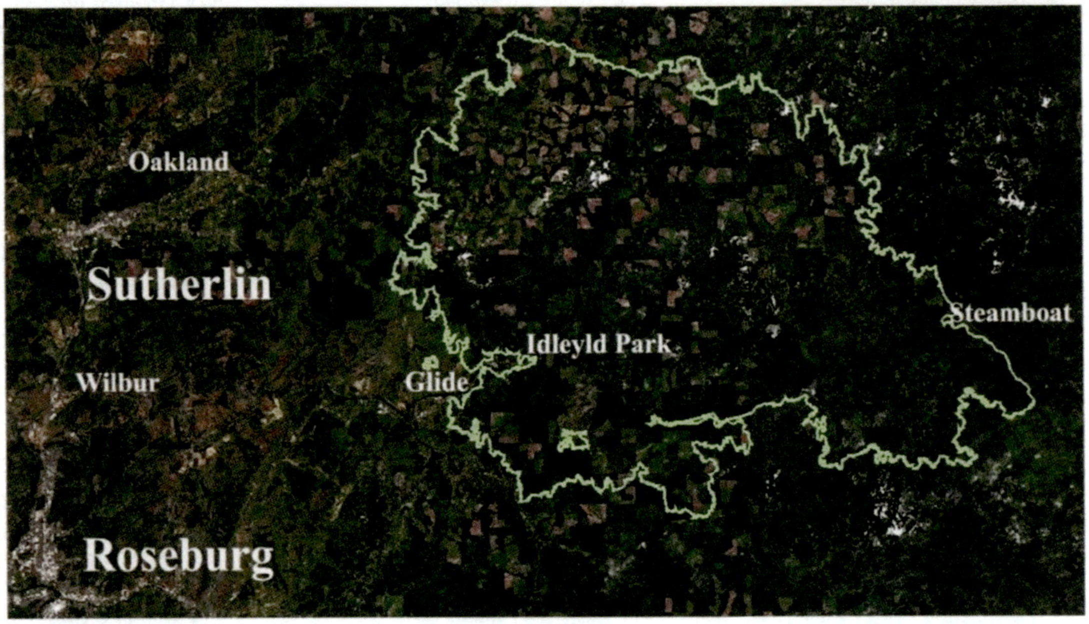

Figure 1. Satellite imagery of North Umpqua River basin and Local communities before Archie Creek Fire, August 2020. Image courtesy Mason, Bruce & Girard.

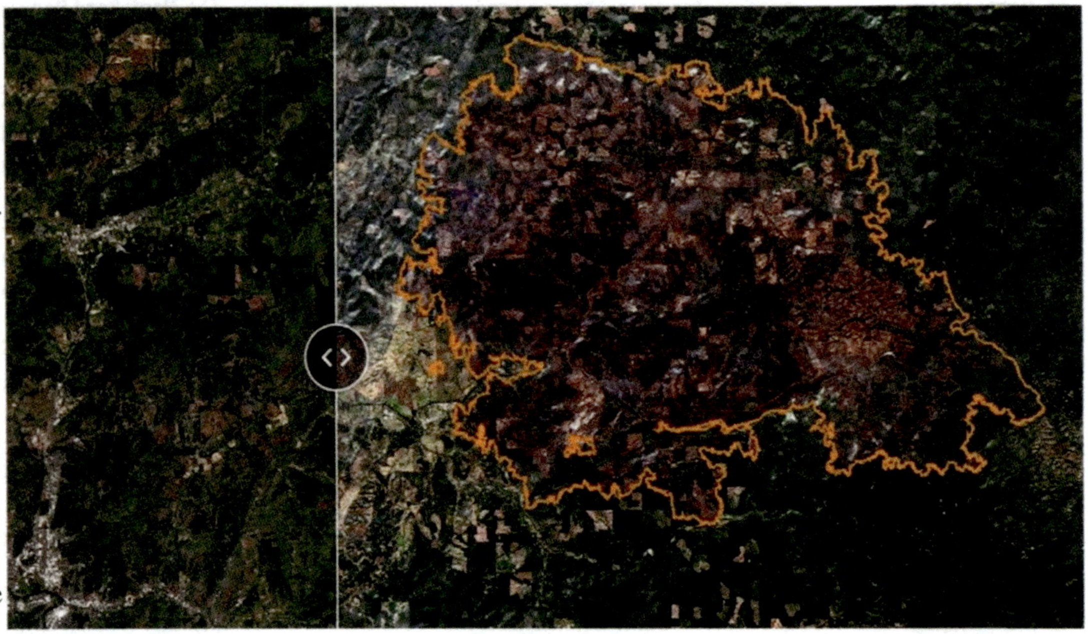

Figure 2. Satellite imagery of North Umpqua River basin following Archie Creek Fire, September 2020. Image courtesy Mason, Bruce & Girard.

Risk of snags remaining on the landscape.

The risk of snags in the environment is well documented. Depending on their size, age, species and distribution, they pose a constant danger of burning in a wildfire. They can also cause death and serious injury when they fall, damage roads and buildings, host harmful insects and diseases and, to many people, are unsightly blemishes on the landscape.

Their greatest danger, though, is their flammability in a wildfire and the deadly and costly destruction that can directly result.

The "Six-Year Jinx" of 1933, 1939, 1945 and 1951 Tillamook Fires demonstrated how catastrophic-scale reburns can take place in western Oregon Douglas fir forests when snags are left in place. It wasn't until the resulting great "Tillamook Burn" had been salvage logged, systematic firebreaks created, and reforested with the help of Oregon schoolchildren that the fires stopped. The salvage logged and planted land is now a State Forest and hasn't burned in more than 70 years.

Another example is the Kalmiopsis Wilderness, in which

no harvesting, road maintenance, or reforestation has taken place for more than 50 years, and which burned and reburned in the 1987 Silver Complex, 2002 Biscuit, 2017 Chetco Bar and 2018 Klondike Fires. It remains to be seen if it will burn again, and when.

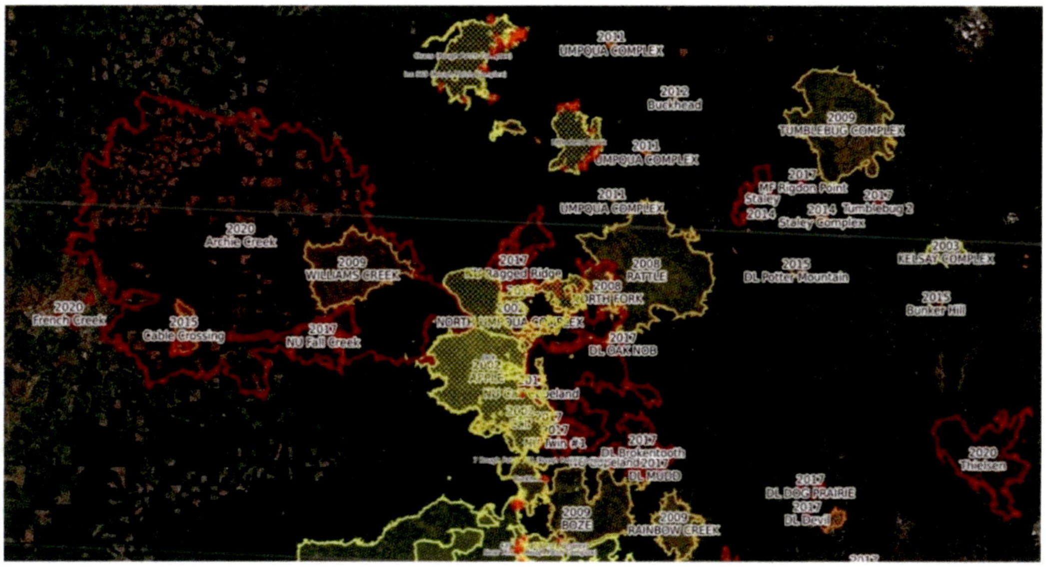

Map 1. Historical North Umpqua River Wildfires, 2002-2021.

In Douglas fir forests the lesson is clear: if widespread snags are not removed or otherwise treated (typically by being felled and broadcast burned), they are almost certain to burn again -- and often hotter and over a larger area than the fires that initially created them.

Prior to the 2009 Williams Creek Fire, the area of the Archie Creek Fire was virtually snag-free. Highly detailed Osborne photographs taken from six separate fire lookouts in 1933 showed a vast landscape of mostly even-aged stands of old-growth and second-growth Douglas fir. Only the Mace Mountain Lookout photos showed a significant number of snags, and they were mostly gone by the 1950s.

Pat Skrip, District Manager, Douglas Forest Protective Association, has estimated it "might take 40 or 50 years" and "about four reburns or other events" for the Archie Creek snags to disappear from the landscape.

Jim Petersen, Evergreen Magazine, has phrased it this way: "Do we cut down our dead and dying trees and put them to good use or do we let Nature burn them down?"

Risk of snags starting and spreading wildfires.

During historical time, virtually every major wildfire in western Oregon has been caused by people or by lightning. Dry lightning strikes typically occur on a seasonal basis in specific locations that can be predicted, but people use fire on a daily basis and fires can be set by careless smokers, campfires, motors, powerlines, arsonists or by other means, nearly everywhere on a year-round basis.

Due to the unusually heavy east winds, massive amounts of dry, pitchy fuels on public lands -- both snags and logs -- and several sources of human ignition, the Labor Day Fires (including Archie Creek) burned westward through the river valleys of the western Cascades like a blowtorch, even moving through the night and killing nearly everything in their paths. The presence of snags in this environment was almost invisible -- everything burned, snags included.

On the North Umpqua, heavy concentrations of snags remained from the 2009 Williams Creek, 2015 Cable Crossing, and 2017 Fall Creek Fires. The Williams Creek snags, in particular, may have contributed to both the initial ignition of the Archie Creek Fire, and also to its rapid spread and intensity (see Map 2).

Burning snags can greatly increase the heat, spread and severity of a fire due to their height and flammability. Such fires can also create and increase their own weather patterns, including high winds, cumulus clouds, lightning, and even rain. Due to their height and with associated winds, firebrands from snags can spread more than a mile in advance of flames, often creating spot fires where they land.

Melvin Thornton has reported "10-foot-long burning limbs" being carried 2,000 feet into the air by wildfire updrafts in western Oregon. He also reports that the 2017 Stouts Creek Fire in the South Umpqua River basin burned so hot and rapidly that its updrafts formed cumulus clouds and lightning -- which strikes were observed starting at least two additional wildfires.

Table 1. Historical North Umpqua River Wildfires, 2002-2021.

Year	*Date*	*Wildfire Name*	*Acres*	*Snags*	*Ownerships*	*X*
2002	0816	Apple	17,600	Yes	USDA NF	A
2008	0813	Rattle	19,800	Yes	USDA Wilderness	L
2009	0912	Boze	10,600	Yes	USDA NF	L
2009	0912	Rainbow	6,100	Yes	USDA NF	L
2009	0728	Williams Creek	8,400	Yes/No	USDA NF/Private	?
2015	0728	Cable Crossing	1,900	Yes/No	USDI BLM/Private	?
2017	0808	Fall Creek	4,800	Yes	USDA NF/USDI BLM	L
2017	0808	Happy Dog	31,400	Yes	USDA NF	L
2017	0808	Twin #1	1,400	Yes	USDA NF	L
2017	0808	Brokentooth	3,700	Yes	USDA NF	L
2020	0908	Thielsen	9,800	Yes	USDA NF	P
2020	0908	Archie Creek	131,500	Yes/No	USDA/USDI/Private	P
2021	0705	Jack Creek	24,000 (+)	Yes/No	USDA NF	P
2021	0730	Chaos	25,400 (+)	Yes	USDA NF	L
2021	0730	Near Minky	4,900 (+)	Yes	USDA NF	L
2021	0730	Buckehead	6,900 (+)	Yes	USDA NF	L
2021	0730	Little Bend Creek	9,000 (+)	Yes	USDA Umpqua NF	L

Acres "(+)" = 2021 wildfires still burning as of September 15, 2021: final acreages unknown.

Snags: "Yes" = Fire starts and/or still remain; "No" = Pre-fire or post-fire salvage and/or felled.

X = Cause: "A" = Arson; "L" = Lightning; "P" = People; "?" = Unknown.

Risk of snags to firefighters.

Attempts to manage or extinguish a wildfire are made many times more dangerous by the presence of burning snags. For this article I conferred with several experts on the topic, a number of whom, including Thornton and Skrip, have had significant professional experience working on the North Umpqua wildfires of the past 20 years.

Why many of these fires were not directly attacked from the air while they were still small has been a point of discussion in recent years. There have been instances in which personnel were too widespread to effectively administer to all fires at one time; others in which seemingly poor decisions were made not to jump; and others in which conditions were simply too dangerous -- including the presence of snags.

According to Chuck Sheley, retired smokejumper and long-time editor of Smokejumper magazine: "Areas with dead snags are not jumped any more. They were definitely dangers during my era. Once you are on the ground and building fire-line, snags are very dangerous."

Map 2. Land Ownerships (private, BLM, US Forest Service) and snag concentrations burned during the first 24 hours of the Archie Creek Fire, September 8, 2020.

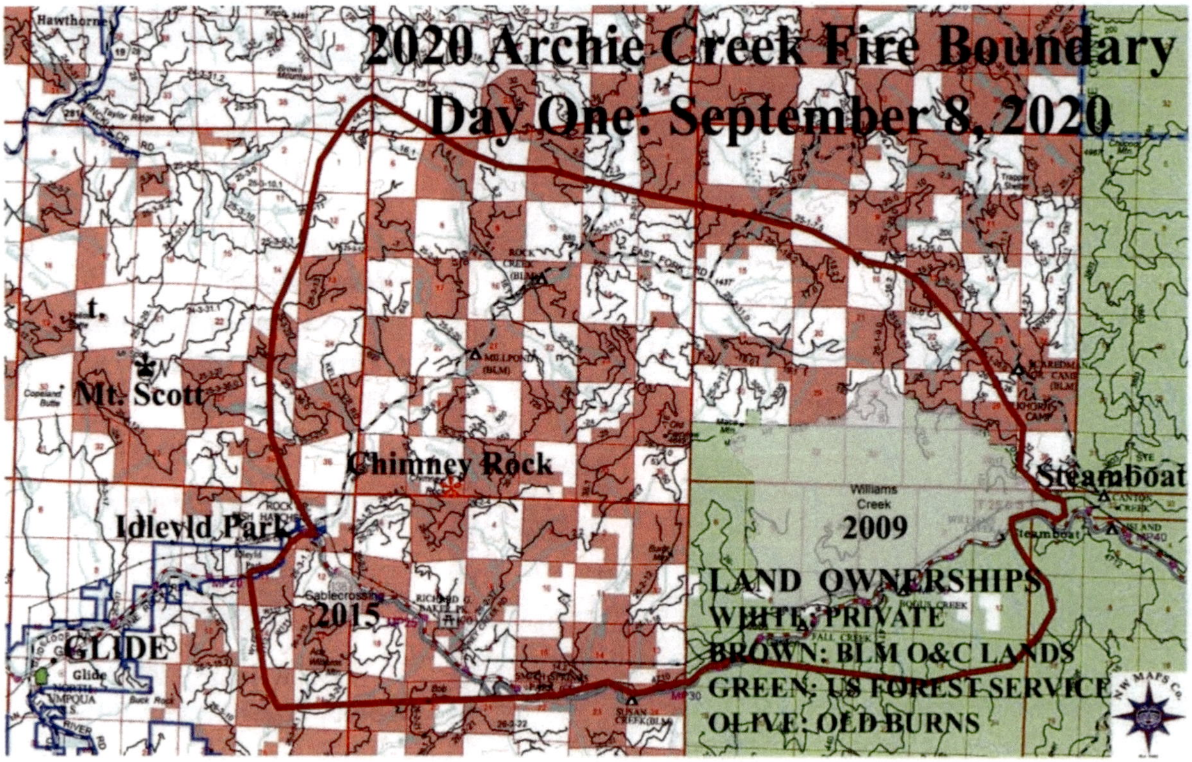

Fire-fighting,

Figure 3. Osborne photograph, view east and southeast from Mount Scott fire lookout, by Albert Arnst and Lester Moe, August 20, 1933.

road clearance, hazard tree removal and other tasks associated with wildfire management and immediate post-fire repair are made significantly more hazardous in the presence of burning snags.

Figure 5 shows the older (and reburned) 2015 Cable Crossing Fire snags have mostly shed all of the limbs, twigs, tops, and needles that are still clearly evident on the more recent 2020 Archie Creek snags. If these latter snags burn in the next year or two they will likely shower burning limbs and embers into the winds, onto other fuels on the ground, and on anyone caught or working beneath them.

Risk of snags to residents.

The 2020 Labor Day Fires illustrated the real risk to rural residents and communities posed by the increasing numbers and expansions of wildfires in western Oregon. More than 4,000 families lost their homes, 11 people died, and everyone

Figure 4. View east and southeast from location of destroyed Mount Scott fire lookout. August 19, 2021. Photograph courtesy of Melvin Thornton, Douglas Forest Protective Association fire detection camera, 1.5x magnification.

Figure 5. Snags remaining from the 2015 Cable Crossing Fire, December 2020. Within the red circle, note blackened trunks and missing bark, limbs, and fine fuels on the reburned snags in the foreground; and the evidence of salvaged snags (bare land) on private property in the background. Also note fine fuels on trees burned in Archie Creek Fire. Drone video-clip by Matt Hill, Douglas Timber Operators.

breathed toxic smoke for days and weeks.

The 2020 Archie Creek and 2021 Jack Fire evacuations clearly affected the communities of Glide, Idleyld Park, Rock Creek and Steamboat. Snags remaining in these locations pose the greatest continuing threats to residents and their properties, both by burning and by falling. The smoke and ash from these fires created serious problems along the I-5 corridor, and particularly the greater communities of Roseburg and Sutherlin.

Figure 6. Streamside buffer snags following salvage logging on Rock Creek tributary, July 19, 2021. Video-clip by McKenzie Peters, NW Maps Co.

Risk of snags to travel and power corridors.

Following the direct threat to people and homes, wildfire risks must be considered in terms of safe evacuations, maintenance of highway transportation, and powerlines needed for water pumps and lights.

"Hazard tree" removals are well-recognized management actions designed to remove dead or dying trees from major travel corridors and electrical lines -- but a more subtle threat is posed by relatively recent regulations regarding "buffer strips" of untouched vegetation along our rivers and their tributaries.

Figure 6, for example, shows a tributary to Rock Creek from a concrete bridge downstream from the "streamside buffer" that remains along its banks. This type of vegetation pattern has been shown to allow wildfire to readily cross streams; to easily uproot, as in the photo, causing siltation; and eventually fall and float downstream toward culverts, railways, bridges and highways.

Figure 5 illustrates the same problem as Figure 6, but at a different scale What will happen to these large snags along the North Umpqua over the next few years? Will they be removed, and possibly sold, or will they be allowed to fall into the river and float downstream during winter floods? Or will they stand long enough to burn again?

Figure 7. Deer hunting was a major recreational use of public forestlands burned in the Archie Creek Fire. These photos were taken by Brook Pfaff, a faller for Lone Rock Timber Co. in September 2020 while clearing right-of-way to company lands in the Rock Creek Basin as the fire was dying down. Sadly, these tragic, painful deaths represent only a fraction of the millions of wild animals and hundreds of species killed in the first 48 hours of the Labor Day Fires on September 8-9, 2020. Animals able to fly or swim to deeper water mostly survived, but mortality of mammals, reptiles, amphibians, insects and fish in shallow waters was nearly 100%. Populations will rebound, but this result could have been prevented.

Risk of snags to private properties.

Private properties are shown as White on Map 2. Most of these properties are industrial timberlands that have already -- within this first year -- been salvage logged and/or even planted, depending on their condition when they burned. Historically, industrial tree plantations have been mostly safe from wildfire, but tree farms in the Rock Creek basin are checkerboarded with square-mile, 640-acre (Brown) BLM Lands which may continue to contain large amounts of snags. In many cases BLM snags will likely fall or spread wildfire on to the adjacent private lands or access roads, which will remain a principal risk to these properties.

Risk of snags to wildlife.

Because snags have largely been salvaged from private lands or are in the process of being treated for reforestation purposes, snags will pose little risk to wildlife on those lands. Rather, the greatest risk to wildlife is on federal lands in which snags and other flammable materials -- such as fallen branches and trees, leaf litter, and exotic weeds -- are allowed to develop and remain (see Figure 7).

Risk of snags to recreation.

The large majority of campsites, trails, and other recreational developments affected by the Archie Creek Fire are on federal lands. Local residents and other visitors to these locations are threatened directly when dead limbs and trees or snags fall, and certainly during times of wildfire. One exception to this general circumstance are the public roads that connect the various O&C Lands. People traveling these routes for recreational purposes -- such as sight-seeing, hunting, camping, or fishing -- may also be endangered or inconvenienced by falling snags or during a wildfire.

Part 1 Summary.

Beginning in 1987, and accelerating after 2002, the number of large-scale fires have increased dramatically on the North Umpqua, and particularly on Umpqua National Forest Land. Following the Archie Creek Fire, the USDA Rapid Assessment Team (RAT) noted:

"Over the past 20 years, 28% of the Umpqua has burned in wildfires, with the total acreage being higher due to several areas being burned two to three times over the past 20 years. Less than 1% of these past fires in total have been salvaged, with the majority of snag loss occurring along roadsides as danger tree mitigation to keep open public access. Therefore, snag abundance at the landscape level will likely be above the 80% tolerance level on the North Umpqua and Diamond Lake Ranger Districts for quite some time."

The important statement in this quote is that "several areas . . . burned two or three times over the past 20 years." Since the RAT report, the North Umpqua 2021 Jack Fire, Chaos Fire and Rough Patch Complex have burned more than 70,000 additional acres on Umpqua National Forest lands and remain un-contained at the time (September 15) of this article.

NEXT ISSUE: The Snags of the North Umpqua: Part 2. Recommendations. NOTE: This article is derived from a report contracted by Douglas Timber Operators with Dr. Zybach and NW Maps Co.

Global Warming and Oregon Wildfires

Part II: History Repeats Itself, Again

By Bob Zybach, Ph.D

Human interference — such as politics and modern techniques to protect forests from blazes — allows tree parasites to lay waste to 90,000 acres of the Deschutes and Willamette national forests.

A natural disaster runs amok

By Theresa Novak
The Stateman Journal

ENVIRONMENT

Tiny worms and voracious beetles that have decimated almost 90,000 acres of trees are almost gone. Many of the dead trees they left behind soon will be gone, too.

For the first time in almost four years, the U.S. Forest Service will sell salvage timber sales on the Sisters and Mcken-zie ranger districts. There, drought and infestation by spruce budworms and bark beetles have killed a vast number of trees in an area bisected by the Santiam Pass.

The damaged area crosses over two federal forests and encompasses some private timberland. Most of the damage is in the Deschutes National Forest, with a smaller portion in the Willamette National Forest near Suttle Lake.

Rich Mathis, who has owned the Suttle Lake resort for three years, said it is about time that the Forest Service did some-

Figure 1. September 3, 1994 front page on Salem's *Statesmen Journal* article describing 90,000 acres of beetle-killed Douglas fir along Highway 20 and the Santiam Pass.

There is a saying that making good decisions is based on experience, and that experience is based on making bad decisions.

I have been writing article/editorials for this magazine on a fairly regular basis for more than 10 years and have been interviewed occasionally by Lars Larson on his radio program for the past 20. In both instances, topics have generally focused on my academic and professional careers regarding wildfires, reforestation, and forest management. And the Elliott.

Another topic has been Global Warming (aka "climate change" and "climate crisis"): based partly on a research paper I delivered to an international group of university researchers in the early 1990s, and mostly on my continuing education on the topic by top scientists and meteorologists as a long-time co-moderator of the Global Warming Realists blog.

This article is focused on the political and scientific predictive history by myself and others regarding the relationship between our changing climate and the great increases in western Oregon forest fires during the past 35 years. It is arranged in three parts with a conclusion:

1) published concerns by me and others regarding increasing risks of dead trees and catastrophic wildfires, from 1991 to 2014;

2) a slightly abbreviated transcript of a live radio interview I did with Lars on these topics in 2014; and

3) the published comments of the governors of Oregon, Washington, and California during and regarding the 2020 Labor Day Fires.

I. 1991-2014 Media Predictions

I began my intermittent research on conifer forest history and climate change in 1991, as Global Warming was just becoming a national concern. On Friday, September 13, I presented my findings to an international conference of climatologists, gathered in Corvallis, on the topic of computerized Global Warming predictive models, carbon cycling, and northern hemisphere conifer forest history.

My presentation (and subsequently published peer-reviewed paper) came to five conclusions, including: "none of the models in current use has demonstrated an ability to make accurate or reliable projections," a fact which I believe continues to be true to the present time.

Figure 1 is from a Fall 1994 front-page newspaper article by Theresa Novak regarding beetle-killed trees along Highway 20 and Santiam Pass.

Figure 2 features a Spring 1994 excerpt from an interview with Jim Petersen that specifically discussed the fire danger from these snags.

Figure 3 is the Fall 2003 B&B Complex Fire cover of this magazine, which featured a photo from the 1994 issue regarding the Highway 20 beetle-kill.

Figure 4 compares the 1994 Novak beetle-kill map with the online 2004 ORWW website map of the B&B Complex Fire.

My Summer 2013 article in this series was about spotted owl habitat and led to an interview with Lars on his radio show that July. Conversely, the transcript of a July 2014 interview with Lars formed the basis of my Fall 2014 article,

"Global Warming and Oregon Wildfire History."

The live interview was conducted on July 29 and included the fact it had been an extremely mild fire season, thus far, in 2014. The following day, July 30, the 25,000-acre Oregon Gulch Fire started (see Table 1).

II. July 2014 Radio Interview

Lars: Welcome back to the Lars Larson Show. You know that we talk about global warming issues on this show on a pretty regular basis and we love to have people on both sides of the question . . . Bob, good to have you back on and thank you very much for taking the time today.

Bob: Well, thank you, Lars. I enjoy it.

Lars: Tom Tidwell is the U.S. Forest Service Chief, and it [news report] was mentioning that he had visited the High Desert Museum and he was interviewed by KTVZ, just so we get all the credit in, and he said because of climate change and biomass, wildfires are different now than they were ten years ago. And a direct quote from Tidwell, '"Our fire seasons today are 60 to 80 days longer, which creates more energy in the system." Tidwell told News Channel 21, "We see more erratic fire behavior." Let's start with the length of the fire season and I want to ask you first. You did your PhD thesis on this subject, right?

Bob: Yes, on the history of [Oregon] Coast Range wildfires . . . so the last 20 or 30 years, I've looked very closely at the history of wildfires in the Pacific Northwest.

Lars: Are the fire seasons two and a half months longer than they used to be?

Bob: No. They are about exactly the same for the last 200 years.

Lars: [Laughs] That's quite a statement. So, for 200 years, we've had fire seasons about the same length. Why do you suppose the U.S. Forest Service Chief, Tom Tidwell, is saying then that they are two to two and a half months longer?

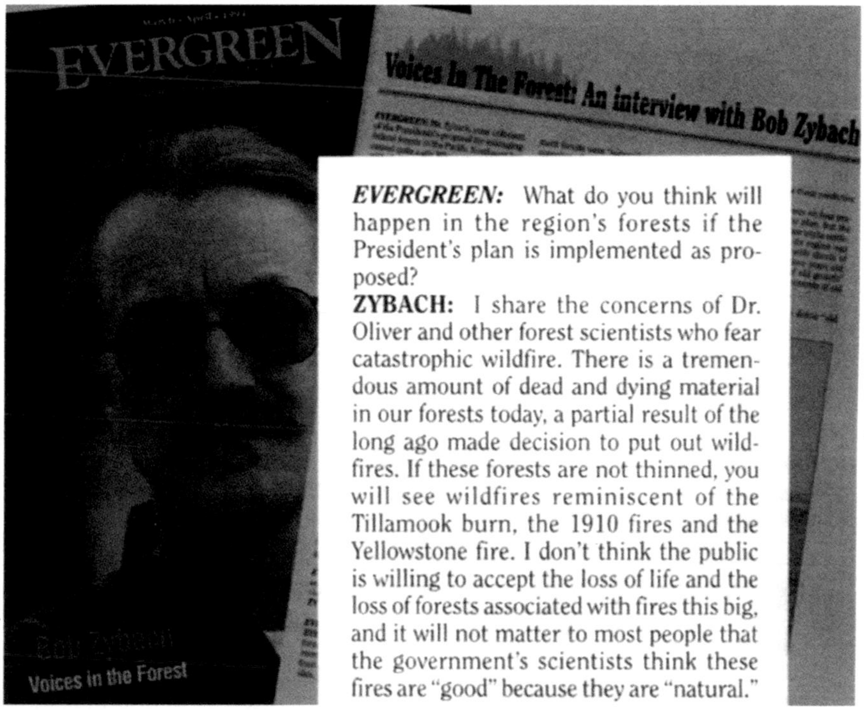

Figure 2. Excerpt from March-April 1994 interview with Jim Petersen, *Evergreen Magazine*, with predictions of catastrophic wildfire on beetle-killed trees on Highway 20 and the Santiam Pass, and fire-killed trees from the 1987 Silver Complex Fire on the Kalmiopsis Wilderness. The 2002 Biscuit Fire reburned the Kalmiopsis and the 2003 B&B Complex Fire burned through Santiam Pass. Both fires were the largest in history for those locations.

Bob: Well maybe because he is following the lead of President Obama, who said last week up in Washington [State] that the wildfires are due to "climate change." I think they have agendas to support. I think it's political. It's not scientific, and that's what bothers me.

Lars: Is it fair to say -- and in fact I had somebody email me -- saying 2014, according to the National Interagency Fire Center, is actually the quietest fire season of the decade so far this year. Is that fair to say?

Bob: That's probably pretty true.

Lars: They say outside of the west coast . . . burn acreage for this year in the United States is the lowest in a decade, less than half of normal, and one of the lowest on record.

Bob: Well, I don't know if it is one of the lowest on record, and I don't know how they are measuring "normal," but the last ten years have been terrible and the last 20 years have been terrible. Actually going back to about 1987. But these fires were predicted more than 20 years ago by Chad Oliver up at University of Washington -- he's at Yale now -- by me, by Jim Peterson, who has Evergreen Magazine --which was nationally distributed -- that the build-up of fuels caused by passive management of our forests was going to result in massive forest fires, and that's exactly what has occurred.

Lars: . . . Is it warmer today than it was ten years ago, or cooler, or about the same temperature?

Bob: I think the records are showing that in the northwest we've got generally cooling temperatures for the last 15 or 17 years. As far as the seasonality of wildfires, they're pretty darn predictable. July and August are the big months. Have been since the 1820s. Sometimes it will start a little bit earlier in June or early July in eastern Oregon, because it dries out sooner. But western Oregon, about now -- late July until mid-August -- is when it gets its worst, and then they usually taper off in September. There's some starts there, and then the rains hit them and that's been the same pattern as long as we've had historical records. It has not changed.

Lars: . . . When Tidwell says something like that, and says the fire seasons are longer today than they used to be . . . You know, small, relatively small changes in temperature are not going to make the difference between say the Coast Range where it's wet and damp and central Oregon where it's bone-dry all summer, is it?

Bob: . . . We have had wildfires here in February and March. It depends on humidity and the wind. You get an east wind and low humidity and the fuels dry out. So fuel build-

Figure 3. Fall 2003 cover to *Oregon Fish & Wildlife Journal*, references 1994 Highway 20 beetle-kill as background to current article on B&B Complex.

ups are a result, in Oregon, of spring rains. That's where you get your vegetation build-up; or management policies that say don't salvage dead material; or let things grow and don't log, say, in forests; or don't graze on the prairies. So it's not a matter of temperature at all. It's humidity and wind and it's fuel. And then it's a source of ignition; and since we have lightning strikes -- not on the Coast Range, but southwest Oregon, eastern Oregon and along the Cascades -- at this time of year, it's predictable that when they strike fuels, they will burn. If we have an east wind and low humidity, they get out of control really quickly. But the temperature doesn't have anything to do with the length of the fire season; and the main thing that controls the annual fuels, the ones that are particularly combustible, are the spring rains -- where you get big, heavy grass build-ups and a lot of woody growth.

Lars: On the subject of fire season, so we have the head of the U.S. Forest Service, saying that we have a longer fire season by 60 to 80 days. I mean that would be a gigantic increase in the length of the fire season, if it was true, but you say we've had basically the same for the last 200 years. Let' talk about the fuel load... Give me your take on that first, if you don't mind, Doc?

Bob: Sure. This probably started -- the fuel load problem that has led into the wildfires -- probably started with the Wilderness Act. A lot of Wilderness areas included timber. The Kalmiopsis in southern Oregon is a good example. In 1987 it burned. There was no salvage because it was a Wilderness area. The Jefferson Wilderness area had beetle-kill in the early 90s. In both those instances, myself and other people said these are going to bum up catastrophically if something isn't done to remove these now dry fuels. Anybody that has used wood heat knows that dry wood burns better than green wood. And then, point of fact, they both did. The B&B Complex burned up almost the exact perimeter of the beetle-kill on the crest of the Cascades in 2003. The previous year, the Biscuit fire burned up the Kalmiopsis again, took out the entire Silver Complex fire that had burned there in 1987, and expanded the boundaries to about 500 thousand acres. So those things were predicted before we had a global warming problem, or about the same time. People, me included, were saying these are time bombs; these fuels are going to go out of control. Since then, we've exacerbated the problem by doing two things: (1) we've created a wildlife habitat for spotted owls, which is burning up for the same reason -- the fuels are building up and there's nothing to control them; and (2) the other reason is lack of salvage. We are not salvaging the areas that have been killed and so we are piling up the dry firewood in those areas and as a result the fires are, as predicted 20 years ago, becoming larger and more destructive.

Lars: And when they talk about trying to ameliorate this through thinning projects . . . That doesn't actually solve the problem, or probably even reduce it, does it?

Bob: Not a bit. I've seen a lot of those projects and they are pretty irritating. They are just busy work. Before I went back to college, I worked for 20 years in reforestation. We did lots of reforestation work throughout the Pacific Northwest and by removing fuels, by putting in road systems, by

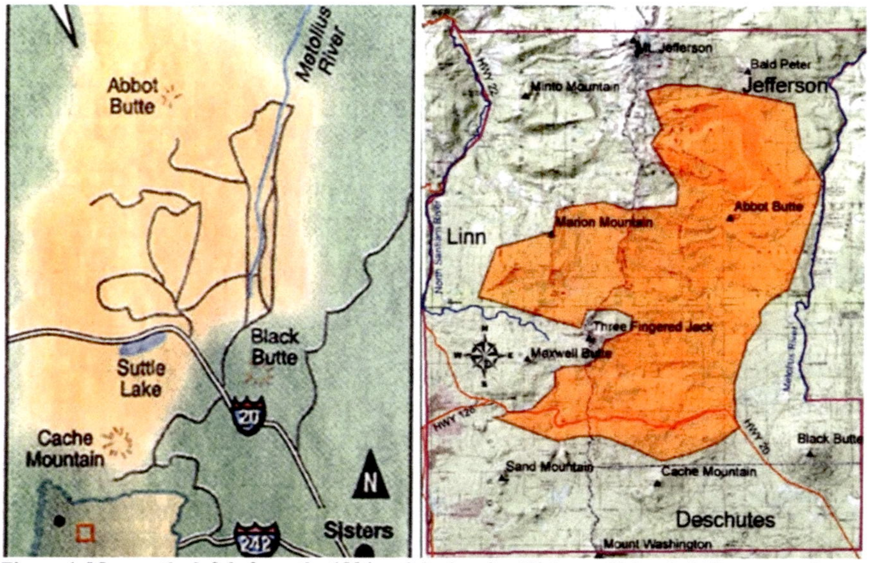

Figure 4. Map on the left is from the 1994 article showing Highway 20 beetle kill; map on the right is from March 2004 ORWW educational website focused on the predicted B&B Complex Fire. Note common Abbot Butte, Metolius River, Black Butte and Cache Mountain landmarks.

using aerial surveillance, and so on, we [people who worked in the woods] were able to keep wildfires at bay from the end of World War II say, until the 1980s, and until the Federal government started putting in passive management policies. The thinning projects they put in are silly, almost. They encourage the remaining trees to grow [greater amounts of fuel]. They are more like precommercial thins or leaf-raking, almost literally, than they are anything that needs to be done. We need to go in and be very aggressive and actively manage the major fuels, which are logs, which also create jobs, which also create income, which also create safer conditions for fish and wildlife and less air pollution and less water pollution.

Lars: And if you do that and, you know, in the process of logging you are knocking off the limbs, you take the limbs and grind them up. I don't know how much commercial value that grind is, but there are markets for some of that stuff and if you take the rest of it and burn it . . . you should be able to do all that and actually improve the forest, rather than just let it build up until it's feet thick and then a fire starts and the fire sits there and burns through this fuel from now till the rains hit.

Bob: And chars the soils and sterilizes it and it sloughs off in the rivers and the whole bit; and creates a lot of carbon dioxide in the process for the people that are worried about that. The way the Indians did it for ten thousand years or so is to have regular prescribed burns and so the fuels never built up and the nutrients that go into the soil or the ash from the plants that [burned] -- the flash fuels mostly -- and those are the forests that are uniformly described as Eden-like by the first explorers and settlers that came to Oregon. The forests were safe. They were beautiful. They were highly productive. Wildlife was productive. So, a combination of logging to get rid of all the fuel build ups of the last 30 and 50 years -- but done along the watershed boundaries and on a landscape scale -- combined with prescribed fires as the Indians did, would pretty much resolve the problem as well as rejuvenating our rural economies and our wildlife populations.

Lars: . . . Dr. Zybach, I appreciate you coming on. We are going to have some links up to some of the material that you have and you always bring a lot of intelligence to the show. Thanks very much.

Bob: Thank you, Lars.

III. 2020 Labor Day Fires Politics

Gavin Newson, Governor, California, September 11: "The debate is over around climate change . . . I'm a little bit exhausted that we have to continue to debate this issue. This is a climate damn

Wildfire Name	Year	Date	Acres	County	Landowner
Big Windy Complex	2013	July 26	26,700	Josephine	BLM
Whiskey Complex	2013	July 26	18,000	Douglas	USFS
Douglas Complex	2013	July 27	48,700	Douglas	BLM
Oregon Gulch	2014	July 30	25,800	Jackson	BLM
Stouts Creek	2015	July 30	26,500	Douglas	Private
Collier Butte	2015	August 2	12,300	Curry	USFS
Chetco Bar	2017	July 12	191,100	Curry	Wilderness
Whitewater	2017	July 23	14,500	Marion	Wilderness
Spruce Lake	2017	July 29	14,500	Jackson	National Park
Happy Dog	2017	August 8	31,400	Douglas	USFS
Jones	2017	August 10	10,100	Lane	USFS
North Umpqua Complex	2017	August 11	43,200	Douglas	USFS
High Cascades Complex	2017	August 13	27,500	Jackson	National Park
Miller Complex	2017	August 14	39,700	Jackson	USFS
Horse Creek Complex	2017	August 21	33,800	Lane	USFS
Horse Prairie	2017	August 26	16,400	Douglas	BLM
Eagle Creek	2017	September 2	48,800	Hood River	USFS
South Umpqua Complex	2018	July 15	28,700	Douglas	USFS
Taylor Creek	2018	July 15	52,800	Josephine	BLM
Klondike	2018	July 15	175,300	Curry	Wilderness
Miles	2018	July 16	54,300	Jackson	USFS
Terwilliger	2018	August 19	11,600	Lane	USFS
Milepost 97	2019	July 24	13,100	Douglas	Tribal
Beachie Creek	2020	August 16	193,600	Marion	Wilderness
Lionshead	2020	August 16	204,500	Marion	Wilderness
Holiday Farm	2020	September 7	173,400	Lane	Private
Slater	2020	September 7	34,00	Josephine	USFS
South Obenchain	2020	September 8	32,700	Jackson	USFS
Riverside	2020	September 8	138,100	Clackamas	USFS
Archie Creek	2020	September 8	131,600	Douglas	USFS
Thielsen	2020	September 8	10,000	Douglas	USFS
Jack Creek	2021	July 5	24,200	Douglas	USFS
Chaos	2021	July 30	28,800	Douglas	USFS
33 Major Fires	**9 Years**	**July-Sept.**	**1,901,734**		

Table 1. Western Oregon large-scale wildfires following topical 2013 and 2014 articles and radio interviews with Lars Larson regarding wildlife habitats and forest fires.

emergency. This is real and it's happening. This is the perfect storm," Newsom, 52, said as he stood on charred, smoky mountain terrain near Oroville in Northern California.

Jay Inslee, Governor, Washington, September 11: "Fires across the Pacific Northwest and California shouldn't be called wildfires, but 'climate fires' . . . This is not an act of God," Inslee said. "This has happened because we have changed the climate of the state of Washington in dramatic ways."

Kate Brown, Governor, Oregon, September 13: "The harsh reality is that we're going to see more of these wildfires. They're hotter, they're more fierce, and obviously much more challenging to tackle. And they are a sign of the changing climate impacts . . . Climate change is here, it's real, and it's like a hammer hitting us in the head . . . And we have to take action . . . This is truly the bellwether for climate change on the West Coast," Brown, a Democrat, said on "Face the Nation." "And this is a wake-up call for all of us that we have got to do everything in our power to tackle climate change."

Conclusions

Between 1952 and 1987, 35 years, only a single wildfire greater than 10,000 acres in size -- the 1966 Oxbow Fire -- occurred in western Oregon. From 1987 until now, even catastrophic-scale wildfires more than 100,000 acres in size have become commonplace, as predicted.

"... we've exacerbated the problem by two things: [1] we've created a wildlife habitat for spotted owls, which is burning up for the same reason the fuels are building up and there's nothing to control them; and [2] the other reason is lack of salvage. We are not salvaging the areas that have been killed and so we are piling up the dry firewood in those areas and as a result the fires are, as predicted 20 years ago, becoming larger and more destructive."

During just the past nine years, from 2013 through 2021, there have been 33 western Oregon wildfires greater than 10,000 acres in size. Of these 33 major wildfires, 30 began and/or burned completely on federal forestlands, and only three began on private or Tribal lands.

The climate is the same for all ownerships -- only management actions have changed. Federal forest managers have harvested less than 5% of the millions of snags remaining from these fires. The remaining 95% will likely burn again, as many already have.

"Those who don't study history are doomed to repeat it. Yet those who *do* study history are doomed to stand by helplessly while everyone else repeats it."

Figure 5. 2022 cartoon by Tom Toro, Portland, Oregon, has gone viral on Internet.

Too many bad decisions have been made by our legislators and public forest managers the past 35 years. We need to learn from these costly errors to make better decisions moving forward.

We can do better. Major wildfires in western Oregon remain predictable and largely preventable. They have had little or nothing to do with "climate change" and nearly everything to do with federal passive management policies. To fix, we just need the legal ability to actively manage our public forests. Again.

HCPs, LSRs, the ESA & Western Oregon Wildfires, 1987 - 2022

By Bob Zybach, Ph.D

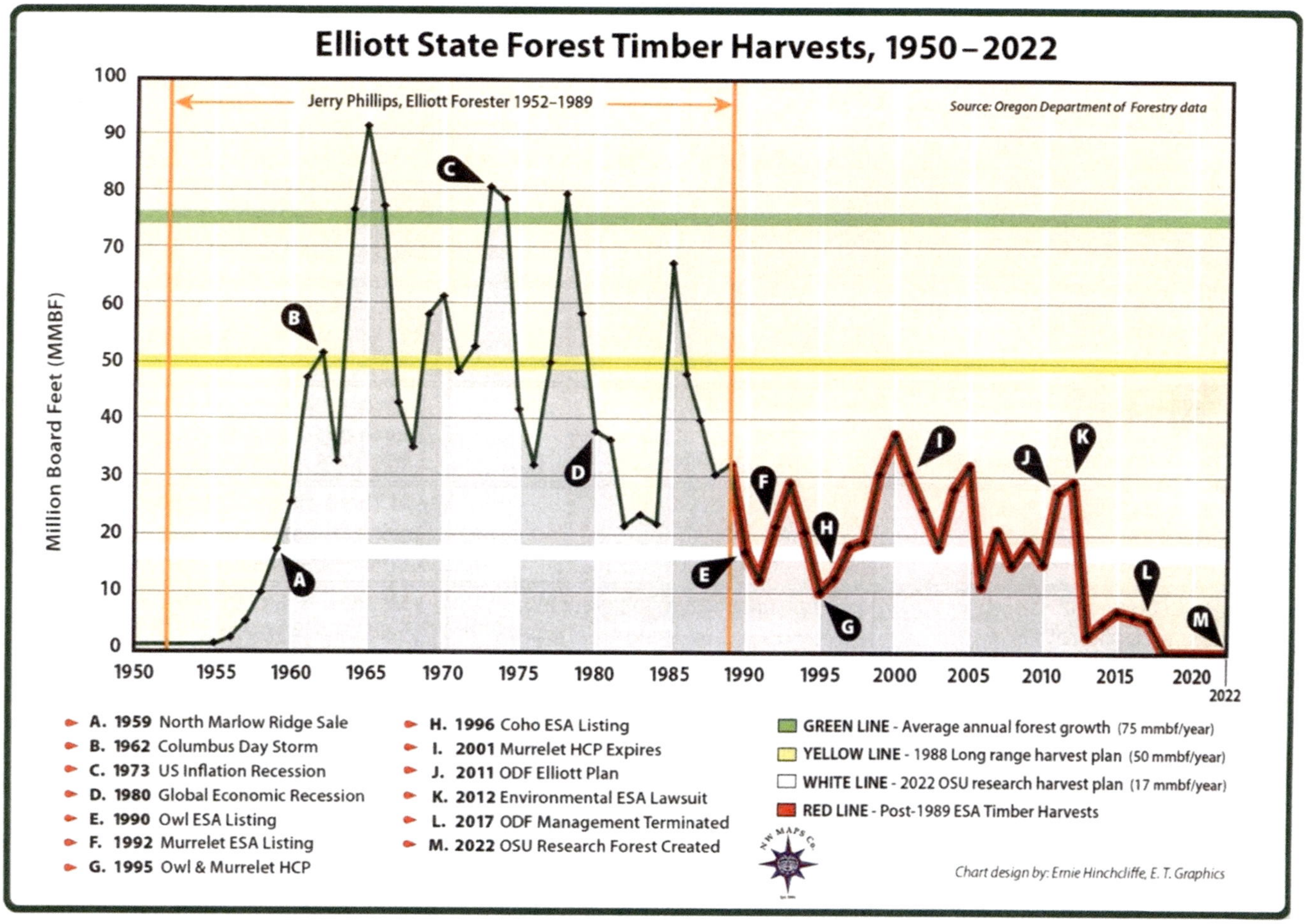

The Green Line represents average annual amount of growth of Elliott State Forest trees; The Yellow Line represents average allowable cut for the Elliott in the 1988 harvest plan; The White Line represents OSU's planned annual harvest of the Elliott w/ no snag salvage; The Area between the Green Line and Red Line represents Elliott fuel increases since 1989.

During the past 35 years, beginning in 1987, western Oregon has experienced the greatest number and extent of catastrophic forest wildfires in its history. Almost all these deadly events took place or began on federal lands managed by the US Forest Service (USFS) or Bureau of Land Management (BLM) under regulations developed from the 1973 Endangered Species Act (ESA) and administered by the US Fish & Wildlife Service (USFWS), National Marine Fisheries Service (NMFS), and/or National Oceanic and Atmospheric Administration (NOAA).

It's the government, so acronyms are obligatory. They are also a form of purposeful obfuscation that help conceal the fact that management of our nation's forests has largely transitioned from the long-term profitable management of trees, wildlife, and recreational facilities by professional foresters, to costly passive management of "critical habitat" by wildlife biologists and computer modelers. Tax-producing loggers and tree planters have largely been replaced by taxpayer-funded agency "ologists," their university modelers, and seasonal wildfire fighters in the process.

This transition took place slowly at first, beginning in the 1960s and creation of Wilderness Areas and the National Environmental Protection Act (NEPA); accelerated in the 1980s with invention of ESA "Habitat Conservation Plans"

(HCPs) and the development of taxpayer funded lawsuits made possible by the Environmental Access to Justice Act (EAJA); and then was almost entirely completed on federal lands in 1994 with the adoption of the Clinton Plan for Northwest Forests (NWFP). Federal access to, and control of, HCPs on state, private, and tribal lands remains a work in progress [see Graph].

From 1952 until 1987 -- also 35 years -- there was only one forest fire more than 10,000 acres in western Oregon, the 1966 43,000-acre Oxbow Fire. Since 1987 there have been at least 36 such fires, with at least eight being more than 100,000 acres. Almost all these fires have taken place on federal lands designated as Wilderness or regulated as NWFP-created "LSRs" (Late-Successional Reserves), "AMRs" (Adaptive Management Reserves), arbitrary streamside buffers ("Riparian Reserves") and other "Congressionally Reserved Areas" -- all now managed, in large part, by USFWS and NMFS for "critical habitat" of select plants and animals.

As the federal government is making the transition from actively managing its roads and forestlands for timber production and recreation to passively managing them for hypothetical habitat requirements of select species, the incidences of large- and catastrophic-scale wildfires have increased 3,600%!

This predictable result was made possible in large part by an endless string of anti-logging lawsuits initiated by a small number of nonprofit environmental organizations.

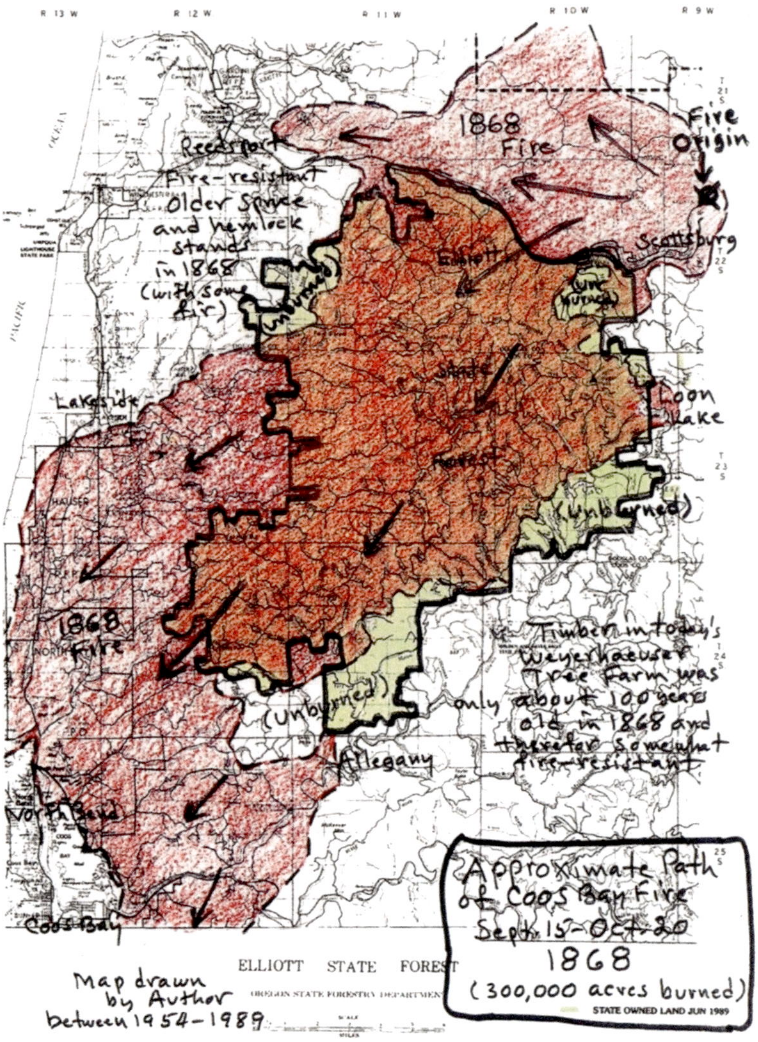

Jerry Phillips' map of the 1868 Coos Fire in relation to present-day Elliott State Forest. This area reburned in 1879, creating the "Big Burn" homesteaded by the Goulds in 1884.

The Center for Biological Diversity (CBD), Portland Audubon, Cascadia Wildlands, and a few others -- with expensive legal teams often unknowingly funded by taxpayers via the EAJA -- have regularly used ESA listings of spotted owls, marbled murrelets, and coho as surrogates to greatly reduce active management of our public forests. Complex NEPA procedures are often the basis for these "successful" filings, and deadly wildfires have often followed. As predicted.

HCPs and The Elliott

I have been writing about forestry, wildfire, and wildlife issues on a regular basis for Oregon Fish & Wildlife Journal for more than 10 years. My previous article in this series regarded the life and career of long-time Elliott State Forest manager, Jerry Phillips. The Elliott history chart at the beginning of this article was also used to illustrate Jerry's great success during his career on the Elliott from 1954 to 1989.

The Elliott was Oregon's first state forest. It was established in 1930 by trading about 70,000 acres of the Siuslaw National Forest located south of the Umpqua River for Common School Fund properties of similar value that were scattered around the State. Previous articles in this series have looked closely at the history of the Elliott and of Oregon's Common School Fund -- which has been managed by law since 1859 for the benefit of Oregon schools by the State Land Board (SLB), composed of the Governor, State Treasurer, and Secretary of the State.

In 1930 the Elliott was mostly covered with young Douglas fir and red alder trees that had seeded in following the catastrophic 1879 "Big Burn" wildfire and several subsequent decades of livestock grazing, firewood gathering, logging, and clearing fires by the Gould's, McClay's, and others who had settled in the area.

In 1955 the Oregon legislature gave the Oregon Department of Forestry (ODF) the responsibility to manage the Elliott more actively as its young timber was growing to merchantable size. The first major logging sale on the Elliott was a stand of mostly 200-year-old Douglas fir on North Marlow Ridge [point A on the graph] in 1959.

Then, on October 12, 1962, everything changed. The Columbus Day Storm [point B] swept over the Elliott without warning. Winds exceeding 150 miles-per-hour blew down 100 million board feet (mmbf) of 70-year-old second-growth timber in a matter of hours. The next several years were devoted to building more than 200 miles of road needed to reach 250 areas filled with toppled "blow-down" -- dead trees that had to be harvested while they still had value and before being infested with beetles or rot. The graph tracks the subsequent annual increases and decreases in harvesting between 1963 and the present. Note the impacts of the 1973 [C] and 1980 [D] recessions on harvest volumes.

In 1988, the Elliott adopted a long-term annual harvest plan average of 50 mmbf, based on the Forest's continued growth and Jerry's continuing success managing natural and human events affecting its development. He retired the following year, as spotted owls were first being discovered on the Forest, and the year after that – 1990 -- the federal government listed the northern spotted owl ["NSO"] as an ESA "threatened species" and the Elliott's 1988 harvest plan was shelved that quick [E].

Elkhorn Ranch, ca. 1894. The Gould family homestead was established near the center of the 1930 Elliott Forest and 1879 "Big Burn" reburned snags from ca. 1775 and 1868 wildfires.

Two years later, in 1992, the marbled murrelet was also listed [F]. Marbled murrelets are birds that spend their entire lives at sea, mostly offshore from Alaska. They can fly more than 60 miles-per-hour and only come ashore some years in early summer to nest a single egg. Small populations of this bird also live in the ocean offshore from Oregon, Washington, and California and occasionally nest on large lateral limbs of Douglas fir, redwood, hemlock, spruce, or bigleaf maple. Adults gather fish daily from the ocean on a high-speed beeline to feed a hatchling until it becomes a fledgling and can fly after four or five weeks -- at which time it immediately also makes a beeline for the ocean, never to return to its nest.

Then, in 1995, spotted owl and marbled murrelet HCPs were developed by the USFWS and adopted by the SLB and ODF for the management of the Elliott [G]. In 1996, coho salmon were also given federal listing [H]. These three animals, spotted owls, marbled murrelets, and coho, have been the basis for most ESA lawsuits filed to stop active management on western Oregon forestlands ever since.

The marbled murrelet HCP expired in 2001 [I] and in 2011 [J] ODF completed its new Elliott Forest Plan, calling for an annual harvest of 40 mmbf -- well below the conservative 1988 plan, but significantly greater than the average harvest since the spotted owl was first listed in 1990.

The following year, 2012 [K], a coalition of environmental groups – Portland's Audubon Society, Eugene's Cascadia Wildlands, and Tucson, Arizona's CBD -- sued the State of Oregon, alleging that the new Elliott plan was illegal. Logging jobs on 28 active sales in State forests were immediately halted as one result.

Rather than contest the court's ruling, the Governor and SLB decided to sell the Elliott at a fraction of its pre-ESA value instead. When this sale was ruled illegal, they transferred ownership to Oregon State University (OSU) for research purposes, terminating ODF's decades-long management role in 2017 [L]. When this transfer was also ruled illegal, the 2022 Oregon State legislature voted to rename the Elliott as the "Elliott State Research Forest" (ESRF), with OSU responsible for its management.

Currently, remaining tasks for the SLB are to "decouple" the Elliott from its legal obligations to the Common School Fund, to develop a new federal HCP, for OSU to develop a functional operating plan, and to locate funding needed to begin proposed research operations.

The current document "puts the cart before the horse" by proposing a major experiment before conducton such an analysis and without developing on the ground familiarity with the property. In addition, the experiment OSU has proposed as badly flawed, compromises developmentof the long-term research potential of the forest and lacks significant revelance to management of Oregon's forests. The proposed experiment violates basic principals essential to production of statistically vaalid and socially convincing outcomes, Futhermore, the focus on Triad, an academic concept related to land allocations at regional scales, has no relevance to pressing forestry issues facing Oregonians. Jerry Franklin and Norm Johnson, November 28, 2020 review of proposed OSU research and management plan for the Elliott.

Top Left: Ground Fuels. Top Right: Ladder Fuels. Bottom: Crown Fuels.
Photographs by McKenzie Peters, NW Maps Co., Mt. Thielsen Trail, August 23, 2020.

In 1994, John Beuter, OSU Forestry economist, was hired by the Oregon Board of Forestry (BOF) to do an economic analysis of the Elliott -- which, despite its new spotted owl status, could still boast 80,000+ forested acres, 2.5+ billion feet of merchantable timber, 550 miles of rocked access roads, and many miles of prime salmon and trout streams. Given recent ODF management history and 1994 ESA restrictions, Beuter concluded: "Selling the Elliott is the only marketing alternative likely to significantly increase net annual income to the CSF [Common School Fund]."

Fifteen years later, in 2009, John Charles, Cascade Policy Institute President and CEO, testified:

"Unfortunately, this recommendation has been consistently ignored by the SLB. Had it been followed in 1995, the CSF today would be worth at least $3 billion, even after the market declines of 2008. Instead, it is only worth about $1 billion. Even by the standards of legislative appropriations, one would think that $2 billion in lost asset value for school funding is something to be concerned about."

Following the 2012 environmental lawsuit, the Elliott began losing even more millions of CSF dollars annually to legal costs, and by 2016 its value had deteriorated to an estimated $220.8 million -- a number determined by Roger Lord, senior appraiser of Mason, Bruce & Girard (MB&G), at the direction of the SLB and developed specifically for the purpose of selling the Forest. This number represented the synthesis of three separate independent Elliott appraisals that had varied from only $192 million to $262 million.

This contracted value was premised on the sale being to the private sector with primary legal constraints to future management only being the Oregon Forest Practices Act (OFPA) and state and federal ESAs. The CSF would somehow be "decoupled" by this sale which, curiously, was not allowed to be even a penny more than $220.8 million. As stated, the sale to a single bidder (of 50 solicited) was ruled illegal and voided.

In 2017 Senator Ted Ferrioli requested that Wayne Giesy and I develop an independent forest management plan for the Elliott that accommodated wildlife habitat requirements while maintaining regular employment for rural workers and continuing to make payments to the CSF. Wayne met several additional times with Senator Ferrioli and had at least two personal meetings with Governor Brown to discuss this proposal.

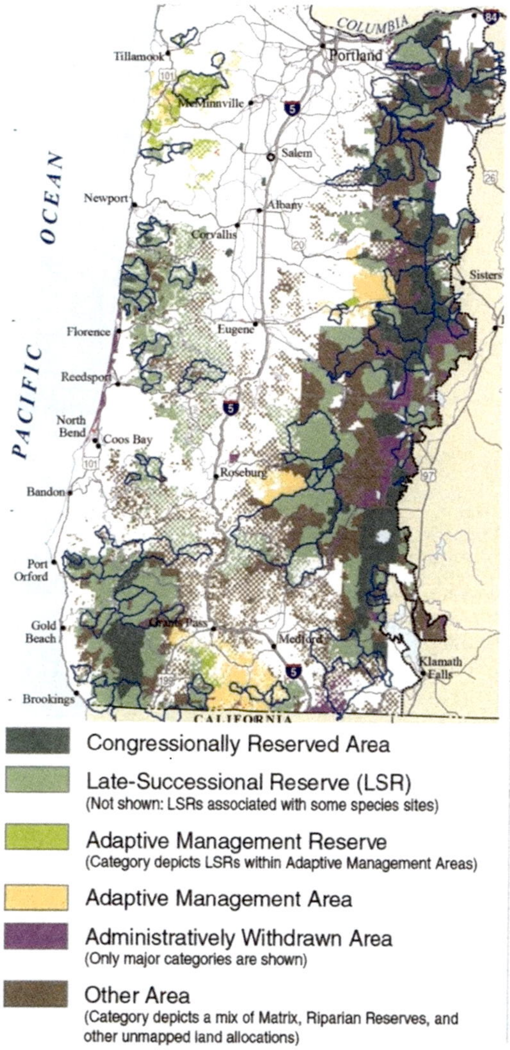

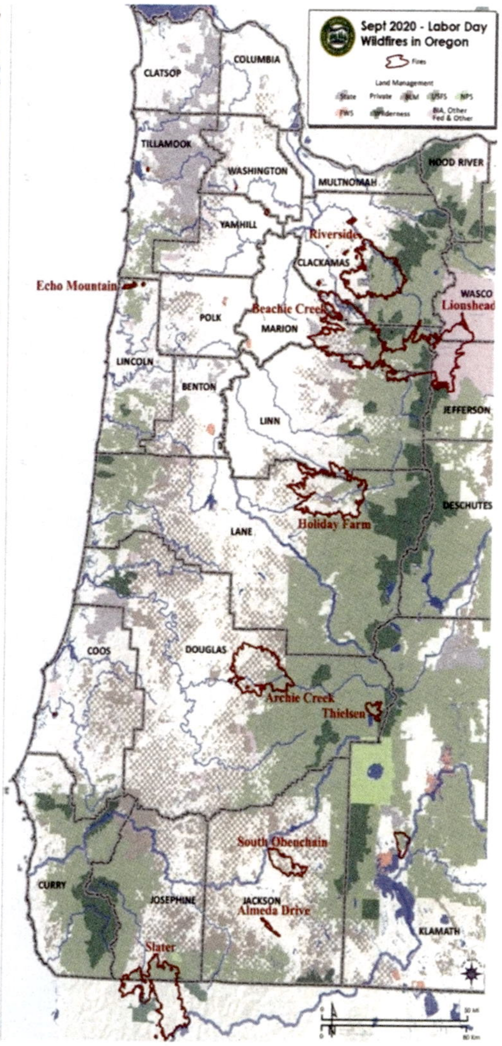

The map and legend on the left are from the 2004 BLM and USFS report on the 1994 Clinton Plan for Northwest Forests. "Congressionally Reserved Areas" are largely Wilderness Areas, created in 1964 and after, and Oregon's only National Park, Crater Lake. The map on the right shows the Labor Day wildfires in 2020: note the correlations between federal Reserves and major wildfires. This pattern is consistent with the entire 1987-2022 time period -- most fires are on federal Reserves.

The resulting "Giesy Plan Alternative" was publicly delivered as requested and has been described in some detail in earlier articles in this series. In part, it adopted the conservative 1988 standard of 50 mmbf/year average annual harvest for an initial 20-year research period. This amount is only about 2/3 of the annual growth of the Elliott and less than 2% of the Forests' estimated 2.5+ billion feet of standing timber, so more volume in older trees of greater value -- dollar and habitat -- would have existed by the end of this two-decade experiment.

At Senator Ferrioli's request, Christine Broniak, then Economist for the Oregon Legislative Revenue Office, estimated potential CSF income based on a 50 mmbf/year average harvest. She used sales prices of Elliott timber for the previous three years of $367.50/mmbf and the "Producer Price Index" to estimate an annual School income of $23 million, or $460 million total for Oregon schools over the proposed 20-year research period.

According to Broniak's math, Oregon schools have hypothetically lost more than $100 million potential income since 2017, and several hundreds of millions more if these numbers are considered for the entire period from 1990. And, despite being publicly and privately requested by the Governor, the Giesy Plan Alternative proposal has never been openly discussed or considered by SLB, Department of State Lands (DSL), Oregon Board of Forestry (BOF), or OSU.

Also in 2017, ODF Forestry Division Chief Liz Dent provided a "conservative estimate of 8.8 direct and indirect jobs for every million feet of harvest" at a public meeting of the SLB. Using that multiplier, the loss of 50 mmbf in timber sales also resulted in the loss of 440 needed rural -- and tax-paying -- jobs when Elliott sales were ended.

In 2022, Lord reduced the MB&G evaluation of the Elliott even more, down to only $99 million! In an August 29 public letter to Geoff Huntington -- now with DSL rather than OSU -- the reasons given for this devaluation were based on the "Market Value" appraisal method used in 2016 vs. the "Investment Value" appraisal method being used 2022.

The principal reasons for this further devaluation of the

Elliott were given as: 1) less acreage available for timber production and 2) less intensive harvesting of those reduced acreages, resulting in 3) lower annual harvest volumes produced at greater costs. Not a good investment. Worth $99 million at most.

HCPs and Oregon Taxpayers

In addition to the never-ending Elliott HCP process, two other major HCP projects are currently taking place in western Oregon -- one for the remaining State Forests still being managed by ODF, and another for the 10 million forested acres owned by private landowners.

Beginning in 2019, Kate Brown organized and convened a select number of environmental organizations, forest landowners, and timber industry representatives together in a series of backroom meetings to develop a "Private Forest Accord" (PFA). This "agreement" would pay landowners money in the form of tax credits instead of logging and reforesting their lands -- and producing jobs and tax revenues -- as they otherwise legally and currently do.

According to the Oregon Wild website (September 21, 2022):

"Ever since the Private Forest Accord agreement between Oregon Wild, our conservation allies, and the logging industry was announced in late October 2021, we've been trying to think of ways to describe the magnitude of the changes that are coming to Oregon's private forest laws.

"Much like the adoption of the Northwest Forest Plan, the passage of the Private Forest Accord does not mean that all issues related to private lands logging are settled. There is still more for communities, Oregon Wild, our partners, and the logging industry to do."

In other words, in addition to devaluing the private properties and putting more people out of work, and having unwitting taxpayers cover the loss in profits with "tax credits" -- the environmental community is also assuming there is "still more . . . to do." Best guess is that lawyers are expected to be involved.

By these general methods, politicians, government bureaucrats, university modelers, and environmental lawyers -- through ESA, NEPA, EAJA, ETC -- have taken almost complete control of our federal forests since 1990. And now they are using USFW and NOAA agencies to gain similar legal access and direct control over private, state, county, and tribal roads, streams, and forestlands. If successful, further rural physical and economic damage and increased severity and extent of local wildfires and directly related losses of homes and wildlife are predicted.

October 2020 unthinned, roadside forest stand on BLM land on Thunder Mountain following Archie Creek Fire. Photo by Matt Hill, Douglas Timber Operators (DTO).

October 2020 photo depicts adjacent roadside stand on BLM land that was thinned and pruned according to prescription, ca. 2015. Photo by Matt Hill, DTO.

HCPs and ODF

In November 2018, the Oregon Board of Forestry (BOF) directed staff to develop an HCP for 17 federally listed species: spotted owls, marbled murrelets, 10 fish (including three separate runs of coho), three salamanders, martens, and the infamous red tree vole. This long-term plan would cover about 640,000 acres of ODF-managed land west of the Cascades, not including the Elliott. If the HCP is approved by USFWS and NOAA, ODF would be assured ESA compliance -- and direct federal access and control -- for 70 years.

For this HCP, "Spotted Owl Habitat" is defined as "includes modeled nesting, roosting, and foraging habitat," and marbled murrelet habitat "includes modeled suitable and highly suitable habitat." Note that these are not scientific documentation of actual birds or habitat -- they are computer "models" of someone's unstated biases and assumptions.

Who, exactly, is determining "suitability" for these imaginary birds? And how? The only "accountability" we get for these printouts is that they are "peer reviewed" by "experts." That's not how science works, and it is certainly not how successful resource management has ever been performed. These are nameless people paid by taxpayers to "model" the directions of committees comprised of professional "experts" and government bureaucrats. What could go wrong?

And the result is a mapping of 275,000 acres of the 640,000 as "critical habitat," with another 164,000 acres of modeled "nesting, roosting and foraging habitat" to be created over the course of the 70-year agreement, meaning that a total of 439,000 -- or 70% -- of our State Forests would be removed from production and put into the hands of federal managers during all of that time.

My concerns with HCPs are: 1) they give federal access and control to private and state lands for decades; 2) they are expensive and costs taxpayers significant money, instead of generating tax revenues; and 3) these government "critical reserves" have a predictable habit of ending in wildfire and killing millions of wildlife supposedly being served.

LSRs and the Clinton Plan

According to Wikipedia (September 19, 2022), the stated purpose of the Endangered Species Act was to protect species and "the ecosystems upon which they depend." The detailed entry contains significant links to CBD and other environmental organizations and writings and makes the following claims:

"About one million species worldwide are currently threatened with extinction. North America alone has lost 3 billion birds since 1970. These significant population declines are a precursor to extinction. Half a million species do not have enough habitat for long-term survival. These species are likely to go extinct in the next few decades without habitat restoration.

"As of January 2019, there are 1,467 total (foreign and domestic) species on the threatened and endangered lists. However, many species have become extinct while on the candidate list or otherwise under consideration for listing. A 2019 report found that FWS faces a backlog of more than 500 species that have been determined to potentially warrant protection."

Certainly, Wikipedia is not an entirely credible source, and some of these statements can be debated, including a general definition of "ecosystem." Instead, ecosystem dependency is defined as: "the present or threatened destruction, modification, or curtailment of its habitat or range" -- apparently a problem of "destroying" or even "modifying" habitat of a species with a known "range."

So now we enter the concept of "critical habitat," which is literally defined as "a habitat area essential to the conservation of a listed species, though the area need not actually be occupied by the species at the time it is designated." In other words, actual occupancy of an area supposedly critical to stop the extinction of a named animal, isn't necessary! A computer model can do the job just fine:

Critical habitat must be designated for all threatened and endangered species, under the Endangered Species Act, with certain specified exceptions. Designations of critical habitats must be based on the "best scientific information available" [BAS, really] and in an open public process within specific timeframes. Unless deemed necessary for the species' continued existence, critical habitat do not include the entire geographical area occupied by a species.

To meet this monumental requirement, the Clinton Plan invented the concept of "Late-Successional Reserves" in order is "to protect and enhance conditions of late-successional and old-growth forest ecosystems, which serve as habitat for late-successional and old-growth related species including the northern spotted owl." With millions of species facing extinction and waiting to be listed, there was only so much "critical habitat" research time and expertise available, so LSRs were created over vast areas of our public forests to account for them all.

Professional predictions that these massive reserves, including Wilderness Areas, LSRs, and streamside buffers would eventually turn into bonfires were ignored, and generally remain unacknowledged to this time. Instead, we get "climate crisis" excuses, WUI acronyms, and massive preventable wildfires almost every year now. We can do better, and have in the past. These are new problems, they are self-inflicted, and they can be fixed. If there is a will.

Conclusions

There is no evidence that HCPs have had any positive effect on targeted spotted owl, marbled murrelet, or coho populations in western Oregon over the past 35 years.

There is documented proof that the adoption of passive approaches to forest management greatly increases accumulations of ground fuels, ladder fuels, and canopy closures, and thereby leads to greater likelihoods of deadly major- and catastrophic-scale wildfires.

Attempts to affect targeted wildlife species by creating Wilderness Areas, Riparian Reserves, LSRs, HCPs and computerized "critical habitat" models in western Oregon during the past 35 years have been at a great cost of billions of dollars, tens of thousands of lost rural jobs, hundreds of thousands of burned forest acreages, thousands of lost homes, millions of killed wildlife, episodes of major air and water pollution, and losses of human life.

There is no evidence that these massive investments have resulted in a single additional owl, murrelet, or salmon.

Eisenhower Was Right Redux: Government Science Is Still Killing Our Trees

By Bob Zybach, Ph.D

Excerpt from President Dwight D. Eisenhoser's Farewell Address, January 17, 1961

"...Akin to, and largely responsible for the sweeping changes in our industrial-military posture, has been the technological revolution during recent decades.

"In this revolution, research has become central; it also becomes more formalized, complex, and costly. A steadily increasing share is conducted for, by, or at the direction of the Federal government.

" Today, the solidtary inventor, tinkering in his shop, has been overshadowed by task forces of scientists in laboratories and testing fields. In the same fashion, the free university, historically the fountainhead of free ideas and scientific discovery, has experienced a revolution in the conduct of research. Partly because of the huge costs involved, a government contract becomes virtually a substitute for intellectual curiosity. For every old blackboard there are now hundreds of new electronic computers.

"The prospect of domination of the nation's scholars by Federal employment. project allocations, and the power of money is ever present and is gravely to be regarded.

"Yet, in holding scientific research and discovery in respect, as we should, we must also be alert to the equal and opposite danger that public policy could itself become captive of a scientific technological elite."

This an update and reconsideration of an article I wrote for this magazine in 2016. Things have not changed much and only become worse so far as the issues discussed at that time: wildfire frequency and extent, wildlife mortality, rural unemployment, private property destruction, air pollution, and aesthetics.

During the past ten years I have written a series of articles and editorials for this magazine that have specifically looked at federal forest management policies, laws, and regulations -- and the so-called "best available science" that is said to be their foundation. These writings have been mostly inspired by the massive changes that have taken place on federal forestlands in Oregon during my lifetime and that have directly resulted in ruined rural economies, broken families, depleted wildlife populations, air and water pollution, degraded landscapes, and millions of acres of dead and dying trees. Almost all of it unnecessary and preventable.

Most of the articles have focused on specific topics, such as wildfire economics, spotted owl habitat, reforestation planning, streamside buffer regulations, cattle grazing along fish bearing streams, or catastrophic wildfire mitigation. This article is essentially a "bottom line" summary of these earlier writings, and less detailed as one result.

The article's title is somewhat misleading in order to be concise and provocative. Eisenhower didn't actually say that "government science" might result in the adverse descriptions given above; he said that government funded scientific research could compromise "intellectual curiosity" and potentially result in misguided policies dictated by a "scientific technological elite."

That is the very process that most concerns me and is typified -- at least in my mind -- by the catastrophic wildfires that have been ravaging our federal forestlands and rural counties the past 35 years.

I Like Ike

The only US President I have ever seen in person was Dwight Eisenhower, with his wife Mamie, traveling in a motorcade from the airport to downtown Portland in 1956. I was a 3rd-grade student at Fernwood Grade School at the time, and our entire class made the short walk to 33rd and Sandy to wave to the President as he passed by. Not much politics in grade school, but "I Like Ike" buttons and stickers were prized possessions of several students with politically involved parents. And everyone knew the jingle:

"Whistle while you work,
[Adlai] Stevenson's a jerk,
Eisenhower has the power,
so whistle while you work."

Eisenhower was a war hero to many of our parents and grandparents and he was a popular peacetime President, largely responsible for our Interstate Highway system and other accomplishments during his tenure.

1961 Eisenhower Speech

I was a 12-year-old 6th-Grader in Baker, Oregon when President Eisenhower delivered his famous "Military-

Industrial Complex" speech on January 17, 1961. This was his televised farewell address to the American people and, even though it has been widely viewed and quoted ever since, I had never read a complete transcript or watched the entire speech until a few years ago.

When I read the complete speech and watched it on YouTube, I was shocked at the accuracy of Eisenhower's warnings and predictions, and particularly as I thought they applied to "science-based" federal forest management policies. His concerns for the future had been transformed into my concerns for the present.

President Dwight Eisenhower and wife Mamie in motorcade following a speech at the Portland Airport, October 19, 1956. Thousands greeted the President at his speech and thousands more lined the streets and cheered as he went by. This photo, by an unknown photographer, was taken at 42nd and Prescott Street, a short distance from 33rd and Sandy along the way to a downtown political rally.

Eisenhower's speech is rightfully famous for its warning to "beware the military-industrial complex" of centralized government agencies and international corporations in control of our food and chemical production, transportation networks, and armament manufacturing. His principal concerns were that Americans might become subjected to a police state ruled by wealthy and influential elites – one potentially in a constant state of warfare because of profits and political power involved. To avert such an undesirable outcome he called for an ever "alert and knowledgeable citizenry."

1933 *Oregonian* newspaper caption: "Looking very much like a head of cauliflower, this column of smoke rising from the terrific [Tillamook] forest fire burning over the coast range was photographed yesterday [August 24] by J.H. Clark of Portland from a plane piloted by Lieutenant B.B. Smith. The head of smoke was between 30,000 and 40,000 feet high, Mr. Clark said."

This warning, of course, was very familiar to me as with most other public-school students in the US during the past 60 years. The part of the speech that was new to me were the following sentences: "The prospect of domination of the nation's scholars by Federal employment, project allocations, and the power of money is ever present and is gravely to be regarded," and "we must also be alert to the equal and opposite danger that public policy could itself become the captive of a scientific technological elite."

To my mind this perfectly described my concerns that agenda-based government research funding and the legal profession were being used to develop expensive computer-based models to dictate policy and expand control over the nation's natural and cultural resources.

On the surface this could possibly be discounted as a "conspiracy theory," but what alternative theory fits any better? The teaching and practice of science has become overtly politicized in the computer age and public policy is being based on the assumptions and manipulations of modelers, "pal reviews," and lawsuits, rather than practical experience and empirical evidence. Or traditional scientific challenges.

1964 Wilderness Act

I was an incoming freshman at Grant High School in Portland when Lyndon Johnson signed the Wilderness Act in September 1964. The first Wilderness areas totaled 9.1 million acres. The action seemed generally popular at the time, but I can't recall a single student or teacher commenting on it and only saw occasional mentions in the news.

This new law recognized "Wilderness" as "an area where the earth and its community of life are untrammeled by man, where man himself is a visitor who does not remain." The Act further defined Wilderness as "an area of undeveloped Federal land retaining its primeval character and influence without permanent improvements or human habitation, which is protected and managed so as to preserve its natural conditions."

This was an expression of an emerging public desire to "scientifically" manage fragile and uncommon lands and wildlife as protective custodians, rather than as active resource managers.

Over time designated Wilderness areas grew dramatically, including millions of acres of former timberlands, ranches, mines, and fisheries. When historians, anthropologists, and archaeologists began to produce strong scientific evidence that people had been "trammeling" many of these areas for thousands of years – including ancient rock carvings, burials, camas meadows, huckleberry fields, obsidian tools, and more recent traces such as orchards, cabins, fences, and wells – the response was to mostly ignore these findings and simply redefine "wilderness."

By the 1980s taxpayer-funded "best available science" (BAS, according to the government) somehow began to support "wilderness" concepts of "natural balance," "niche ecology," and "non-declining, even-flow, naturally-functioning ecosystems," in which humans were mostly seen as pathogens and their observed presence only degraded and threatened "the natural environment."

Today, four federal agencies -- USDA Forest Service, USDA National Park Service, USDI Bureau of Land Management (BLM), and US Fish and Wildlife Service -- administer 759 Wildernesses encompassing 109,754,604

The author, planting trees in the old Yaquina Burn near Eddyville, Oregon in 1981. Photo by Bruce Fraser, Phoenix Communications.

acres, totaling more than 5% of all US lands.

1966 Endangered Species Preservation Act

I was an 18-year-old High School graduate with my first tree-planting job, near Diamond Lake in Douglas County, when Congress passed the 1966 Endangered Species Preservation Act. Whooping cranes, grizzly bears, timber wolves, and bald eagles had been in the news for several years and most people – myself included – thought it was probably a good idea to "conserve, protect, restore, and propagate certain species of native fish and wildlife."

Year	*Wildfire Name*	*Acres*	*County 1*	*Ownerships & Agencies*
1951	Hubbard Creek	15,600	**Douglas**	**Private**
1951	Vincent Creek	28,200	**Lane**	**USDA Siuslaw NF**
1951	Sardine Creek	17,500	**Marion**	Private
1951	HeeHee	5,000	Marion	**USDA Willamette NF**
1951	Tillamook IV	32,700	**Tillamook**	County
1966	**Oxbow**	**42,900**	**Lane**	**USDI BLM O&C Lands**
1987	Bland Mountain I	10,300	Douglas	USDI BLM O&C Lands
1987	Douglas Complex I	30,000	Douglas	USDI BLM O&C Lands
1987	Silver Complex	96,000	**Josephine**	**USDI Kalmiopsis Wilderness**
1991	Warner Creek	9,000	Lane	USDA Willamette NF
1992	East Evans	10,100	**Jackson**	USDI BLM O&C Lands
1994	Hull Mountain	8,000	Jackson	Private
1996	Spring	16,400	Douglas	**USDA Umpqua NF**
1996	Charlton	10,400	Lane	**USDI Waldo Lake Wilderness**
2002	Apple	17,600	Douglas	USDA Umpqua NF
2002	Tiller Complex	69,800	Douglas	USDA Umpqua NF
2002	Timbered Rock	27,400	Jackson	USDI BLM O&C Lands
2002	Biscuit	500,000	Josephine	USDI Kalmiopsis Wilderness
2003	B&B Complex	90,800	**Linn**	**USDI Mount Jefferson Wilderness**
2004	Bland Mountain II	4,700	Douglas	USDI O&C Lands
2005	Blossom	14,800	**Curry**	**USDA Wild Rogue Wilderness**
2005	Deer Creek	1,548	Josephine	**USDA Siskiyou NF**
2008	Rattle	19,800	Douglas	USDA Umpqua NF
2008	Middle Fork	21,100	Jackson	**USDI Sky Lakes Wilderness**
2009	Boze	10,600	Douglas	USDA Umpqua NF
2009	Rainbow Creek	6,100	Douglas	USDA Umpqua NF
2009	Williams Creek	8,400	Douglas	USDA Umpqua NF
2009	Tumblebug	14,600	Lane	USDA Willamette NF
2010	Oak Flat	7,500	Josephine	USDA Siskiyou NF
2011	Dollar Lake	6,300	**Clackamas**	**USDI Mount Hood Wilderness**
2013	Douglas Complex II	48,700	Douglas	USDI BLM O&C Lands
2013	Whiskey Complex	18,000	Douglas	USDA Umpqua NF
2013	Big Windy Complex	26,700	Josephine	USDI BLM O&C Lands
2014	Oregon Gulch	25,800	Jackson	USDI BLM O&C Lands
2015	Collier Butte	12,300	Curry	USDA Siskiyou NF
2015	Stouts Creek	26,500	Douglas	Private
2015	National Creek Complex	21,000	**Klamath**	**USDI Crater Lake National Park**

Table 1. Major Western Oregon Wildfires, 1951-2016 (66 Years).

Year	*Wildfire Name*	*Acres*	*County 1*	*Ownerships & Agencies*
2017	Chetco Bar	191,000	**Curry**	**USDI Kalmiopsis Wilderness**
2017	Happy Dog	31,400	**Douglas**	**USDA Umpqua NF**
2017	Horse Prairie	16,400	Douglas	USDA Umpqua NF
2017	North Umpqua Complex	43,200	Douglas	USDA Umpqua NF
2017	Eagle Creek	48,800	**Hood River**	**USDA Mount Hood NF**
2017	High Cascades Complex	27,500	**Jackson**	**USDI Crater Lake National Park**
2017	Miller Complex	39,700	Jackson	**USDA Siskiyou NF**
2017	Spruce Lake	14,500	Jackson	USDI Crater Lake National Park
2017	Horse Creek Complex	33,800	**Lane**	**USDA Willamette NF**
2017	Jones	10,100	Lane	USDA Willamette NF
2017	Whitewater	14,500	**Marion**	**USDI Mount Jefferson Wilderness**
2018	Klondike	175,300	Curry	USDI Kalmiopsis Wilderness
2018	South Umpqua Complex	28,700	Douglas	USDA Umpqua NF
2018	Miles	54,300	Jackson	**USDA Rogue River NF**
2018	Taylor Creek	52,800	**Josephine**	**USDI BLM O&C Lands**
2018	Terwilliger	11,600	Lane	USDA Willamette NF
2019	Milepost 97	13,100	Douglas	**Tribal**
2020	Riverside	138,100	**Clackamas**	USDA Mount Hood NF
2020	Archie Creek	131,600	Douglas	USDA Umpqua NF
2020	Thielsen	10,000	Douglas	USDA Umpqua NF
2020	South Enchain	32,700	Jackson	USDA Rogue River NF
2020	Slater	34,000	Josephine	USDA Siskiyou NF
2020	Holiday Farm	173,400	Lane	**Private**
2020	Beachie Creek	193,600	Marion	**USDI Opal Creek Wilderness**
2020	Lionshead	204,500	Marion	USDI Mount Jefferson Wilderness
2021	Chaos Complex	28,800	Douglas	USDA Umpqua NF
2021	Jack Creek	24,200	Douglas	USDA Umpqua NF
2022	Rum Creek	21,300	Josephine	USDI BLM O&C Lands
2022	Cedar Creek	127,300	Lane	**USDI Waldo Lake Wilderness**

Table 2. Major Western Oregon Wildfires, 2017-2022 (6 Years).

In March 1967 the original "endangered species" list included 14 mammals, 36 birds, three reptiles, three amphibians and had a budget of $15 million per year. The money was intended to purchase "habitat" for animals on the list for the new National Wildlife Refuge System.

In 1973, with strong support of the Nixon administration, Congress almost unanimously passed a completely rewritten Endangered Species Act (ESA). The new law distinguished threatened from endangered species, allowed listing of a species in danger in just part of its range, allowed listing of plants and invertebrates, authorized unlimited funds for species protection, and made it illegal to kill, harm, or otherwise "take" a listed species. In effect, according to the Center for Biological Diversity, "the law made endangered species protection the highest priority of government."

Today, in 2023, citing laws and regulations requiring "the best scientific data available," there are more than 2,370 plants and animals listed by the ESA, of which only 46 (1.9%) have ever been determined "recovered" -- including most recently a wolf, a flying squirrel, and four wildflowers. The annual budget is now more than $300 million per year.

1970 EPA Clean Air Act

In January 1970 I formed a reforestation business with two friends and began performing tree-planting projects for BLM in Coos County; I was a young business owner with a wife and new baby to support. The previous month President Richard Nixon had signed the National Environmental Policy Act (NEPA) into law, declaring: "the 1970s [will be] a historic period when, by conscious choice, [we] transform our land into what we want it to become" in his State of the Union Address.

It is interesting to consider what Nixon "wanted our land to become" in 1969 compared to what we now have, and how people now might "want the land to become" in the future. In Nixon's time loggers were still "getting out the cut," and often entire hillsides were "slicked off" during logging operations. There was no such thing as a "riparian buffer strip," and logging and tree planting were routinely conducted to the very edge of rivers and streams. Fishing, hunting, hiking, and camping were generally excellent and new roads and trails were going everywhere the fish and game – and logs and wildfires – were.

The Clean Air Act was signed into law by Nixon during 1970 and, on December 2nd, he created the Environmental Protection Agency (EPA) to enforce these new regulations. This Agency almost single-handedly created an entirely new industry – one that has continued to grow and thrive to this time: the practice of "environmental law" by hundreds of EPA lawyers responding to the hundreds of legal suits filed by hundreds of lawyers paid by taxpayer-subsidized non-profit organizations.

Dr. Ben Stout and snags remaining from 2002 B&B Complex Fire near western shore of Round Lake. May 15, 2004 photo by author.

Computerized "models" of "habitat" and "climate" and "fire return intervals" and the number of government scientists, technicians and lawyers needed to develop, promote, and defend these new regulatory terms accelerated rapidly following the creation of EPA. New policies, laws, and regulations sprang – and continue to spring – from their wake. The agency now has an annual budget in excess of $8 billion and employs more than 17,000 people, of which more than half are highly paid engineers, scientists, lawyers, and "policy analysts."

1994 Clinton Northwest Forest Plan

In 1981 my reforestation business was identified as one of the "500 fastest growing businesses in the US" by Inc. Magazine -- the only such business (#332) so identified from Oregon that year. I now had two kids, a small fleet of trucks, more than 30 regular employees, a new log home of my own design, a few hundred acres of manicured timberland, and represented the reforestation industry at a Congressional hearing about Oregon Wildernesses, chaired by Senator Mark Hatfield. A few years later we were broke, my wife and I separated, and the land and property sold -- just like hundreds of other rural family-owned forestry businesses in the western US at that time.

When I subsequently attended forestry classes at Oregon State University in the late 1980s and 1990s there was a lot of interest in such concepts as "preserving old-growth forests," "maintaining spotted owl habitat," and "riparian enhancement." And virtually nothing on active forest management, timber sales, or reforestation planning.

These concepts were typically rationalized by unfounded theories of "steady-state ecosystems" and idealistic descriptions of such circumstances as "non-declining, even-flow, naturally functioning" forests and grasslands. It didn't seem to matter that such conditions had never actually been observed in nature, measured, or documented – only that, for some reason for some people, they were strongly desired.

In 1994 these theoretical "ideals" were integrated into President Clinton's "North- west Forest Plan" as a "focus on scientifically sound, ecologically credible, and legally responsible strategies and implementation." Since its adoption, the plan has been directly associated with worsening of rural forestry-based economies; a significant increase in the number and severity of large-scale wildfires; and a documented decrease in several desired native plant and animal species, including deer, elk, spotted owls, oak, and huckleberries.

Table 1 is how it appeared in the 2016 version of this article, documenting the size and location of major Oregon wildfires in western Oregon since 1950. The following year included the final "Six Year Jinx" of Tillamook Fires, beginning in 1933 and recurring in 1939 and 1945, before ending in 1951. In 1962 I filled my first deer tag in the Tillamook Burn, and in 1963 my High School Biology Club planted seedlings there, among thousands of other Oregon students who had similarly helped reforest the area over the years.

The Oxbow Fire was dying down as I was planting my first commercial seedlings in the Fall of 1966. For the next 20 years of my reforestation career, there were no major forest fires in western Oregon. None. The lands were being actively managed, rural economies, schools, parks, and roads were all in good shape, no homeless people to speak of, and then the Wildernesses began catching fire.

These fires were soon followed by the passively managed Clinton Plan "reserves" bursting into flames -- as publicly predicted by me and several others. These predictions -- based on experience, documentation, and traditional scientific analysis -- were ignored in favor of the government and university modelers focused on "old-growth," "critical habitat," "streamside buffers," and other no-logging set asides. When these measures failed -- as predicted -- the excuse became "Global Warming," and even more job security for the modelers, politicians, lawyers, and "non-profit" environmental organizations.

Look at the numbers beginning in 1952 and continuing until 1986 -- 35 years -- and compare them with the wildfires that took place from 1987 until 2016 (30 years). Now look at Table 2, which lists the major western Oregon wildfires of 2017 through 2022 -- only six years! Next, compare the catastrophic 1951 wildfires to those of 2017 and 2020. Note that more than 90% of these fires are taking place on federal lands -- which has the same climate as private, state, and tribal lands -- as predicted.

The stark difference in recent fire histories has been identified and discussed in several of my earlier articles as "active management vs. passive management." Active management is typified by such activities as road and trail maintenance, vegetation and wildlife management, reforestation planning, and/or recreational developments – all of which took place on federal forestlands from 1951-1986 and continue to take place on private and industrial forests today. Passive management is typified by Wilderness creations, roadless areas, spotted owl habitat, and riparian buffers, as outlined, and characterized by decisions to do very little or nothing until these areas are in flames.

2016 (and 2023): Science vs. Modeling

Eisenhower was right. In the years since his farewell address, taxpayer-funded research and related policies and litigation have been accompanied by massive numbers of expensive and restrictive federal regulations requiring huge bureaucracies and thousands of lawyers to enforce -- and all apparently based on findings and desires of a like-minded corps of government-funded computer-centric modelers and technicians: the "scientific technological elite."

Most of the current policies, laws, and regulations governing our federal, state, tribal, private, and municipal lands, waters, and resources are now based upon the dictates of these elites. One problem of many, is that a significant number of these policies are also based on disproven theories that are inherently racist and strongly biased against past cultures and current populations; a fact that has gone largely unnoticed and unchallenged. Why that is might be hard to explain, but these biases are obviously based more on personal values and political realities than on empirical findings or actual research. The added fact that this growing fountain of environmental laws and regulations is claimed to be -- and is being taught as -- "science-based" is even more troubling.

The practice and teaching of science has been seriously compromised during this process, and our rural economies and environments have been significantly damaged as one result. The same argument can be made regarding damages to our native plant and animal populations, the degraded quality of our scenery, air, and waters -- as well as the teaching and practice of science itself.

Better late than never. My opinion remains that now is still a good time to return to traditional scientific methods to guide our resource management policies -- and to also develop a common faith and understanding of legitimate experience, earned knowledge, and to better value "intellectual curiosity" over "government contracts," as Eisenhower counseled.

It only seems right that we leave our descendants something similar to the wonderful conditions we were given by our own ancestors on the land.

Conversations with Lars: Summer of '23 Smoke & Fire

By Bob Zybach, PhD.

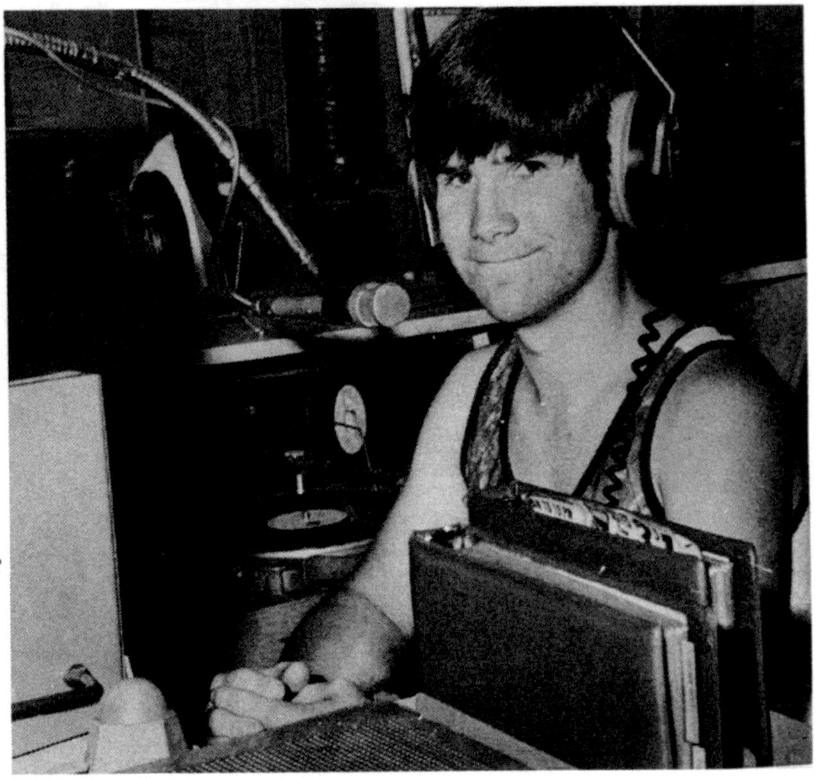

Hey Lars! *Larson was only a 16-year-old at Tillamook High School when he began his broadcasting career, seven days a week at "Mighty 1590" KTIL. He was hired and mentored by Mildred Davy, station manager, and paid $1.85 an hour working school nights from 10PM to midnight, before closing the station. Saturdays, 7PM to midnight; and 5AM to 10AM on Sundays. After graduation he briefly attended the University of Oregon and began picking up broadcasting jobs in Eugene and Portland. In 1997 he started hosting the Lars Larson Show at KXL in Portland, which expanded to the Northwest Network in 2000, and then syndiacated nationally in 2003. Today the Northwest show is carried live from noon until 3PM weekdays to an audience of a million listeners, followed by his national show from 3PM weekdays to 6PM to more than 100 affiliate stations and between four and five million listeners.*

I've never met Lars Larson in person, but my first radio interview with him was about 20 years ago as I was finishing my graduate degree at Oregon State University. The questions likely had something to do with the Biscuit Fire at that time, or the Donato Study, which was in the news.

Since then we have had many more conversations on air, with discussions mostly focused on spotted owls, wildfires, forestry, or the Elliott State Forest. These are subjects of particular interest to me, and it's always a pleasure talking to Lars -- usually in nine-minute increments between commercials -- given his own knowledge of these topics.

Because of Lars' close familiarity with forestry, Northwest history, wildfires, and wildlife, his interviews are more like discussions or conversations than typical interviews. For that reason I decided to use the transcript from our recorded July 28, 2014 talk as the basis for an article/editorial in this series. The topic was the ever-increasing severity, frequency, and extent of Oregon catastrophic wildfires -- as I had been clearly predicting for many years -- and "climate change" as a possible cause. The article appeared in the Fall 2014 issue of this magazine, titled "Global Warming and Oregon Wildfire History," and was generally well received.

Unfortunately, my predictions of increasingly common and greater wildfires, due to the passive and ever-increasing federal regulations guiding the management of our public forests, continued to prove accurate. The fire seasons of 2015, 2017, and 2020 were deadly to both people and wildlife, killed millions of acres of trees, including much old-growth, and took place in only six years. It was decided that an updating and reprinting of the 2014 article and transcript would be timely, so "Part II. History Repeats Itself, Again" of the "Global Warming and Oregon Fire History" saga was published in the Spring 2022 issue.

This summer I had wildfire-related conversations with Lars on two of his shows. In July we discussed the smoke from Canadian wildfires polluting US air, and in August the topic was the deadly fire in Hawaii. Audio recordings of both interviews were critically well received by several national and regional experts in wildfire management and mitigation, and I decided to resurrect the 2014 format for this article.

The past weekend, in early September, was the first time I have ever talked with Lars and not had our conversation broadcast to tens or hundreds of thousands of his listeners. How come he knows so much about forestry and wildfires and wildlife, anyway? I've listened to him for nearly 40 years on television and the radio and had some idea, but not much.

And though I have never met Lars in person, I learned I must have met his father, Waldo Larson, when we were both extras on the set of "Paint Your Wagon" in the summer of 1968. The movie was shot in the Wallowa Mountains near Baker Oregon and starred Lee Marvin, Jean Seberg, and Clint Eastwood. There were about 60 or 80 of us "atmosphere" on set daily and in costume for months, but we were different generations: very familiar and friendly on sight; separate groups of associates during breaks and when otherwise not working.

Waldo was about my father's age, and they both served in the Navy in the South Pacific during WW II, Waldo in submarines and Dad aboard ship. Also, Waldo continued in the Navy after the war ended and served a full 20 years. In 1954 he married retired Navy Lieutenant Junior Grade Harriet Mills, who had served as a nurse during the war. The couple was married in San Francisco, but had relocated to Taipei, Taiwan by the time Lars was born in 1959. Navy families move a lot.

Waldo retired from the service in 1961 and the young family moved to Montana so he could attend forestry school. To make ends meet they lived in a small cabin with an outhouse while Waldo attended classes at the University of Montana, and in a lookout tower for three summers, where he was seasonally employed by the US Forest Service. After graduation and a short stint as a National Park Service Ranger at Mt. Glacier and then Mt. Rainier, the family moved from Somes Bar to Happy Camp, and then Dorris in the deep woods of northern California, where Lars attended grade schools and Waldo worked as a forester.

From Dorris, on the Oregon border, the family moved to nearby Klamath Falls. On the night of December 31, 1969 their car was struck by a drunken driver. Lars was mostly okay, but Harriet had been killed and his father was seriously injured. The family returned to Montana and lived with relatives while he healed.

A year or so later, when he was able, Waldo returned to Klamath Falls, where he worked in a sawmill and Lars finished the 6th Grade. Then he got a job as an Oregon State Park Ranger, based in Tillamook, and Lars went to Junior High School, and then High School, where he became a skilled debater, a proud Cheesemaker, and began his broadcasting career.

July 7, 2023: SMOKE

Lars: Welcome back to the Lars Larson Show, it's a pleasure to be with you, and I'm always glad to get to your phone calls and your emails. On this First Amendment Friday, we celebrate your first amendment rights of free speech, free expression, and the right to associate with anyone you want to associate with.

Now, I want to ask you about this. You've seen what happened when wildfire smoke got into New York City and all of a sudden, the elites were breathing that orange air, and you've seen the same kind of thing happen this year in Seattle, they're breathing smoky air as well. And in recent history in the Pacific Northwest, we've seen plenty of occasions where the smoke went on for weeks and weeks and weeks.

Waldo Larson served in the Navy for 20 years before getting a degree in Forestry and completing his career as a Park Ranger at Cape Lookout State Park near Tillamook. His passion and hobby, was as an extra in Hollywood films, beginning in 1968 and continuing after his retirement. By 1993 he had appeared in 22 different films, including; "Silverado," "Pale Rider," "Red Dawn," and "Stand By Me." This photo is from his role as a character also named "Waldo" in the Johnny Depp movie, "Benny and Joon." The movie-Waldo was a clean-up man in Benny's (Aidan Quinn) garage, where he was interviewed on-set by the Seattle Times in 1993: "I've been in a bunch of westerns where I play an old sheepherder or an old drunken cowboy. I'm real good at falling off bar stools. That's one of my specialties."

So, the perfect guy to talk to, the man who usually doesn't get interviewed by the mainstream media, is the guy who knows the subject very well. Dr. Bob Zybach, forest scientist, president of Northwest Maps Company, and the author of The Great Fires: Indian Burning and Catastrophic Forest Fire Patterns of the Oregon Coast Range, 1491 to 1951. It covers quite a span. Dr. Zybach, welcome back.

Bob: Well, thank you, Lars. It's good to be back.

Lars: I'm glad to have you here because I want you to prepare my audience. I know this summer we're likely to get even more smoke in the Pacific Northwest. We're going to have fires. We've gone from the 30-year period you talk about frequently from I think the mid-fifties to the mid-eighties where we had essentially no large fires in the forest. Thirty years of no large fires at all, and now routinely half a million acres burn in Oregon. Half a million acres I think, on an average year burn in Washington, and we'll expect to have fires this year as well. And I know to a fair certainty, the people in charge are going to say, "Yep, it's all evidence of global warming." Help prepare them with some answers for those people who say those things to them.

Bob: Well, it's all due to fuels and weather. Global warming hasn't happened here, so it can't be global warming. We've got the same weather we've had for centuries. Fire season is the problem. East winds are the danger. So with the Labor Day Fires three years ago, we had east winds in early September, so we had massive fires.

The real problem is managing the fuels. From '52 to '87 we had one major fire, on the Smith River in 1966. It was 40,000 acres. So that's 30 some years with one major fire. It doesn't even compare to the Labor Day Fires -- on one day, where close to a million acres burned. The Coast Range doesn't get lightning; Southwest Oregon gets lightning but doesn't have a lot of people; and the Western Cascades gets lightning and has a lot of people. So once we get a heavy east wind, assuming we do, ignition can come from lightning or people and large fires are the result. And largely because of the massive fuel buildups on federal lands over the last 35 years.

Lars: Dr. Zybach, there's one thing I hear the media do constantly and they say, "Wildfires get worse during hot times." Is there anything about a day being either 80-degrees or 105-degrees that makes a difference in terms of fire?

Bob: It's the east wind. You can have an 80-degree or 105-degree fire; maybe at 105 degrees, depending on fuel moisture, you could have a cleaner burn and easier to control by that measure, but it depends. The fire will create its own wind, will create its own weather. They can even create thunderstorms, the big fires. But an east wind is the constant element that goes with all the major fires in Western Oregon over the last few hundred years.

Lars: So when you see these fires, you've studied this subject, you studied 500 years of it, 1491 to 1951. Are there ways to get on top of this problem where we could prevent these fires instead of merely trying to put them out every year and usually succeeding only to the extent that we

Larson estimates that he shot "about a dozen bucks and a dozen bull elk with a rifle," but this is his first deer with a compound bow. The photo is from late summer 2018 near Terrebonne, with Smith Rock in the background. Since then he has tagged a second buck with a bow.

contain them to half a million acres; instead of maybe 10 or 20,000 acres a year on an average year in that period you documented from the fifties to the eighties?

Bob: Yes, and it would be the same thing. It would be active management. Right now, the Forest Service and BLM are planning to leave all the snags and large woody debris. Jerry Franklin says a sign of a healthy forest is a lot of dead trees. That'd be like saying a sign of a healthy city is a lot of dead people. It doesn't make sense. That's not healthy. It's a fuel and it's dangerous.

So we used to harvest snags, dead trees, focus on it. We used to maintain the roads and trails and keep them open. We used to have local employment where local people that knew the roads and the land and the animals were the ones that were doing the logging and the tree planting and so on. And so we didn't have fires. So we know how to mitigate these fires and that's why they're so predictable.

Like myself and others in the early nineties said, "If we create these LSRs and other government acronyms, we're going to have massive wildfires and they're going to kill wildlife and some people, destroy homes, and it's going to be at a cost to rural communities that lose the work associated with forest management." So we know how to fix the problem. We just don't.

Lars: I'm just curious, do you have any insights as to why the people who followed you in forest science, I mean you've been in forest science for decades, why the people who are now coming into it seem to think that forests that burn on a regular basis or forests with lots of dead fuel on the forest floor are a healthy forest? Why the change?

Bob: Well, indoctrination. Eisenhower warned us that the government will get big computers and put independent scientists out of business. And essentially, if you cut to the chase, it's anti-logging activists. A lot of people in the eighties and nineties thought that clearcutting was an evil. And so they picked spotted owls and marbled murrelets and coho as animals that they claimed -- erroneously, still erroneously -- were harmed directly by clearcutting, and got the lawyers in Washington DC and the lawyers on the ground to pursue that. And they've been very, very successful. And wildfires are the predictable result.

Lars: I guess what frustrates me Dr. Zybach, is that most of the forests that have burned are not on the private lands. It's not Weyerhaeuser's, not the private companies like Hampton. It's the public lands, the state lands, the federal lands, the BLM Lands, county lands, and that we're passing up a gigantic opportunity to harvest billions, hundreds of billions of dollars' worth of trees and make jobs and everything else. And you could do it at a rate that's sustainable, meaning you just keep planting trees and you never run out of trees. Am I wrong to think of it that way?

New York, New York, July 7, 2023. Photo by Anthony Quintano, Creative Commons.

Bob: No. We demonstrated that was right. They probably would do things in the seventies and eighties that we wouldn't do now, but it's a matter of modifying what works rather than abandoning it. And that's what's happened. And that's another argument why this isn't global warming. Is it only global warming on federal lands where all the fires are happening and not happening on Indian lands and private lands? Well, it's global. So that argument is out the window. It's total mismanagement. It was predicted from the beginning, and they're making the situation worse by not salvaging the snags and dead wood that has been the results of these fires since the 1990s, but certainly since 2017 and 2020.

August 14, 2023: FIRE

Lars: Welcome back to the Lars Larson Show. It's a Tuesday. It's the Radio Northwest Network, and it's my pleasure to be with you. And now we have the deadliest fire in US history in Maui and northwest communities under evacuation orders from wildfires. What has put us in this spot and how do we get out of it?

There are actually some sensible solutions and even some that might even involve making money and making jobs. We'll talk about it with Dr. Bob Zybach, who's a forest scientist. You've heard him on the program before, president of Northwest Maps Company and author of The Great Fires: Indian Burning and the Catastrophic Forest Fire Patterns of the Oregon Coast Range. Doctor, welcome back.

Bob: Hi, Lars. Good to be back.

Lars: I want to get your take initially about what happened in Maui because we're now starting to see not just where the blame may go to power lines or other conditions like that, but almost everybody on the left politically says, oh, this is all about climate change. This is something that's come on us because human beings use too many fossil fuels. Any truth to that?

Bob: None. It has got nothing to do with climate change. Everything to do with housing, exotic weeds, in the case of Hawaii; which is similar to Paradise and similar to the Almeda Drive fire in that weeds and housing that were very close together formed the primary fuels and in all three cases were deadly. People died because of the speed in which the fire moved.

Lars: Well, and weeds in the case -- I know when people hear weeds, they say well, everybody has weeds, but is it worse in places like Maui? Because as I understand, that used to be a big area for growing sugar cane and then sugar cane has gone away to a large extent, and as a result, there's a bunch of land that's not very well tended but it could be, couldn't it?

Bob: Yeah, and it's the same thing. Almeda Drive was weeds. They created a "Greenway" and it grew up in blackberries, and those blackberries are real volatile when they die and form a canopy, and that's what happened there; that and a lot of trailer houses and a lot of weeds. In Hawaii it was weeds that grew up in the agricultural areas that had been abandoned or converted to housing and then the housing is essentially dead trees. It's dried lumber that's built the houses, and if they're close together, there's no way to make them "Fire Safe." Each house is fuel for the adjacent house.

Lars: And is part of the problem that in Lahaina especially, they called the place historic. It had a lot of buildings that went back before modern building codes. If they had said, well, even without building codes, we have to do something to keep these houses if one catches fire from spreading to the next one. This was all foreseeable and preventable. Am I wrong?

Bob: No, that's exactly right. When we find -- a wind will whip up in different directions. Here in Western Oregon, it's from the east, and in Hawaii, I'm guessing it might be from any direction -- but when weeds and fuels, volatile fuels, are adjacent to flammable buildings where people live, it's a risk and we're seeing the results of that risk.

Lars: Now, what about the Northwest communities? We've got a bunch of communities that they've gone to evacuation, mandatory evacuation. Are those also evidence that we're not managing the forest and the wildlands very well and we could be?

Bob: We're doing a terrible job there. These fires were predictable for the last 30 years. If we look at the Flat Fire right now, the heat isn't a real problem. It means fuels burn cleaner and faster. But if a wind comes up, if a Chetco Wind comes up, an east wind comes up, we're asking for another Silver Complex or Chetco Bar Fire, just a real disaster. And the way we can tell that is the Flat Fire. It's well contained at about 33,000 acres that used to be called a major fire 20 years ago. Now it's got to be a hundred thousand acres to be a major fire, but you look at the photos and it's completely surrounded by snags from earlier fires.

Maui, Hawaii, August 10, 2023. Photo by Patrick T. Fallon, Getty Images.

So these fires are just fueling future fires just like the Tillamook Fires in the 1930s and '40s did. And it wasn't until we removed the snags, took out the fuels and actively managed that land that we were able to create Tillamook Forest, and we've got the same problem in Curry County. They're just allowing the snags to remain in place and fuel the next fire, and it's been going on since 1987.

Lars: I'm talking to Dr. Bob Zybach, who's a forest scientist and the president of Northwest Maps. The other thing is they don't just allow the snags to stay there. Correct me if I'm wrong, but the "Greeny Groups" say, "No, you will not go in and salvage log. No, you will not take down those snags. That's part of Mother Nature. We have to leave it there." And they insist on leaving it there and not clearing the fuel and not replanting. I've seen that demand a number of times. And they could actually do that, and maybe even make the money to pay the cost of doing the replanting, couldn't they?

Bob: Sure. We did that for 30 or 40 years. When we studied the Tillamook, we figured out if we salvage this material, we're making money, we're paying taxes, we're training people, we're keeping access roads open. And so from '52 until '87, we had one major fire, one fire in excess of 10,000 acres in Western Oregon. Now we have a fire that big right now that we're holding and calling contained, or 56% contained. So it's a problem that's become exacerbated through mismanagement of our federal lands specifically, but now that's being transferred over to our private and state lands as well. The Elliott State Forest, they have no plan to harvest any snags, so it's just asking for a disaster at some future point.

Lars: And by the way, the Elliott State Forest is an issue you've been dogging for a long time. Correct me if I'm wrong, that forest used to spin off tens of millions of dollars for schools in Oregon. It now spins off no money, it just costs money, and yet nothing has changed. The trees are still there to make the money from. It's just the attitude about logging has changed and they say we don't want to log. We don't want to make money from the Elliot Forest. We just want to set it up, I guess, to burn down at some point.

Bob: It's really sad. There's 400 jobs that should be going on there. It produced tens of millions for our schools. It was created as our first State Forest specifically to fund our schools, and now taxpayers are paying for lawyers and "researchers" -- I'm going to put quotes around it -- to let it lie fallow and develop into an increasing wildfire risk. It's burned severely several times, and now they've got somebody counting tree rings and saying that that's their fire history; whereas Jerry Phillips and myself and other people have developed really detailed fire histories showing exactly what the risk is to the communities, to the west of Elliott, and yet they're growing snags there just like on the federal lands.

Lars: Dr. Zybach, you may have better numbers than I, but the last time I looked up the average fires in the last 10 or 20 years, it comes out to about a half a million acres in Oregon a year and a half a million in Washington. I believe that's the number. And then when we have a big fire year, we get a million acres burned. That is compared to the 35 years you talked about where we had one fire bigger than 10,000 acres and nothing else in the way of major fires for 35 years.

Bob: Yep. The Labor Day fires, we had nearly a million acres burn at one point. The Tillamook Fires every six years were burning a hundred thousand to 300,000 acres, and those were the lone events, but from '52 to '87, one fire in excess of 10,000 acres. The others were put out and maintained, managed, and lots of jobs and our communities and schools and parks were doing very well.

Lars: Well, from your lips to the politician's ears because somebody has to change the thinking on this, otherwise it won't be global warming. It'll just be our fire . . . our forests are going to burn on a regular basis. Every year, we're going to lose a half a million acres, and then we're not going to go in and clear it off and get the fuels out and salvage what we can. We're just going to set it up for the next fire.

Dr. Bob Zybach is a forest scientist. We're glad to call on him. I wish we had more forest scientists like that right now, and maybe if we did and some of that thinking in politicians in the northwest, we'd stop these fires. Instead, it's just become a regular occurrence and they always blame it on global warming.

"... we've exacerbated the problem by two things: (1) we've created a wildlife habitat for spotted owls, which is burning up for the same reason, the fuels are building up and there's nothing to control them; and (2) the other reason is lack of salvage. We are not salvaging the areas that have been killed and so we are piling up the dry firewood in those areas and as a result the fires are, as predicted 20 years ago, becoming larger and more destructive."

This quote is excerpted from The Lars Larson Show, July 28, 2014 transcript and was highlighted in the 2014 and 2022 "Global Warming and Oregon Fire History" articles described in the text.

WESTERN OREGON WILFIRES: AN ANTHOLOGY

Made in United States
Orlando, FL
23 June 2026

83294422R00088